THE GOLDEN VERSES OF PYTHAGORAS

Seek this wisdom by doing service, by strong search, by questions, and by humility; the wise who see the truth will communicate it unto thee, and knowing which thou shalt never fall into error.

SHRI KRISHNA

SACRED TEXTS

This series of fresh renderings of sacred texts from the world's chief religions is an inspiring testament to the universality of the human spirit. Each text is accompanied by an instructive essay as an aid to reflection. In the ancient world, before the proliferation of print, seekers of wisdom thought it a great privilege to learn a text, and sought oral instruction from a Teacher in the quest for enlightenment. The scriptures of all traditions are guides to the attainment of serene continuity of consciousness through the practice of self-study, self-transcendence and self-regeneration in daily life.

SACRED TEXTS

CHANTS FOR CONTEMPLATION by Guru Nanak

THE DIAMOND SUTRA (from the Final Teachings of the Buddha)

THE GATHAS OF ZARATHUSTRA — The Sermons of Zoroaster

IN THE BEGINNING — The Mystical Meaning of Genesis

THE GOLDEN VERSES OF PYTHAGORAS
(with the commentary of Hierocles)

RETURN TO SHIVA (From the *Yoga Vasishtha Maharamayana*)

THE SEALS OF WISDOM — The Essence of Islamic Mysticism
by Ibn al-'Arabi

SELF-PURIFICATION (Jaina Sutra)

THE GOLDEN VERSES
OF PYTHAGORAS

The *Golden Verses* constitute the oldest extant Pythagorean teaching. Delivered by Pythagoras (6th century B.C.) and codified by his disciples, they convey the essential core of the exemplary way of life he initiated at Crotona. Hierocles composed his commentary on the *Golden Verses* at some time between A.D. 415 and 450 while he occupied the successor's chair of the Alexandrian Academy, the direct offshoot of the Platonic Academy in Athens. The translation of the commentary included here is adapted from that of N. Rowe.

THE GOLDEN VERSES
OF
PYTHAGORAS

WITH THE COMMENTARY
OF HIEROCLES

Adapted from the translation of

N. ROWE

CONCORD GROVE PRESS

1983

CONCORD GROVE PRESS
1407 Chapala St., Santa Barbara, CA 93101, U.S.A.

First Printing: August 12, 1983

ISBN 0-88695-009-0
10 9 8 7 6

CONTENTS

Imagine a spark or seed of Fire planted in the womb of unlimited Darkness. By the self-propagating power of light spreading outwards from this centre, a spherical realm of order and form and colour is won from the dominion of Night. This is the universe, the Cosmos, extending from earth at the centre to the sphere of the fixed stars. Between centre and circumference, the seven known planets (including sun and moon) are set, each like a jewel in its ring, in material orbits which carry them round; and these are spaced according to the intervals of a musical scale, the celestial harmonia. . . . Such, in its earliest form, is the harmony of the spheres. 'The whole heaven', said the Pythagoreans, 'is harmony and number' — number because the essence of harmony lies, not in the sound, but in the numerical proportions, and these (I think we may add) constitute the soul of Nature, which thus, like the human soul, is itself a harmony. So the moral world is interfused with the physical. The harmony of heaven is perfect; but its counterpart in human souls is marred with imperfection and discord. This is what we call vice or evil. The attainment of that purity which is to release the soul at last from the wheel of incarnation, may now be construed as the reproduction, in the individual, of the cosmic harmony — the divine order of the world. Herein lies the secret of the power of music over the soul.

F. M. CORNFORD

Osiris, Krishna, Buddha, Christ, will be shown as different means for one and the same royal highway of final bliss — Nirvana. Mystical Christianity teaches *self*-redemption through one's own seventh principle, the liberated Paramatma, called by the one Christ, by the others Buddha; this is equivalent to regeneration, or rebirth in spirit, and it therefore expounds just the same truth as the Nirvana of Buddhism. All of us have to get rid of our own Ego, the illusory, apparent self, to recognise our true Self, in a transcendental divine life.

THE MAHA CHOHAN

PYTHAGORAS AND HIS SCHOOL

Pythagoras was revered in India as Pitar Guru, Father and Teacher, and as Yavanacharya, the Ionian philosopher. He was known by other names in ancient Egypt where he spent twenty years in preparation before, at the age of fifty-six,* he founded the School at Crotona in Magna Graecia, with great deliberation and in accord with the wisdom and the vision of the mighty Brotherhood he represented. He taught an entire emerging community, seeking four hundred pure souls who might constitute a small brotherhood for the sake of making that polis a city of souls in search of wisdom in harmony with the larger fellowship of man. His School was based upon the most stringent rules for admission, including a probation lasting five years and a requirement of total silence in the presence of those in the assembly who had been longer in the school. He initiated those who had passed all the preliminary trials, making themselves channels for the divine fount of omniscience, towards which he always pointed and upon which he enjoined an absolute, reverential silence.

For Pythagoras, philosophy was a purgation of the mind and emotions so that the pure light of the immortal soul may freely shine through the limited vestures common to all men. The purification must begin by preparatory reverence — becoming truly worthy of relationship through silent worship of the immortal gods with the transcendental order which holds everything in the universe in a divine harmony. This order could be seen in the heavens and be studied with the help of geometry and mathematics of the most archetypal form. Through the honouring of heroes and peers, profound reverence may emerge for the whole of life when seen in the context of a vast universe. Pythagoras was the first to use the word 'cosmos'. The universe is a cosmos, not a chaos. It represents the majesty of a vast intelligible ordering of immense magnitudes through the application of an overbrooding architectonic principle in a rounded but boundless perspective. It is bounded in time and space but unbounded in its peripheral transition to the realm of the potential. Apprehending this, a person begins to deepen his or her feeling for the mystery of life

* Pythagoras lived from about 582 B.C. to about 507 B.C. Crotona was founded around 536 B.C.

and all the multitudinous forms of matter, and thereby comes to have a true and rectified respect for those forces that are ceaselessly at work, even in the simplest acts such as the handling of objects. The person who is thus prepared would naturally honour the noble heroes, the forerunners of every race and every civilization who, though they are imperfect individuals, are yet capable of elevating the moral tone of human culture. Throughout history their name is legion. Anyone who has thought about these matters fosters a view of human nature that is enormously expansive, and comes to see human beings in terms of opportunities, not limitations, in terms of powers and possibilities rather than handicaps. Then, according to the Pythagorean teaching in the *Golden Verses,* any person can come to show fearlessness in relation to fate, having already acquired a mature self-respect that is rooted in an understanding and a reverence for all of life.

Self-respect means here very much more than in current usage and in our ordinary languages. It is the key to what is said in the *Golden Verses* about proper self-examination, which is an activity very different from offering a confessional before a priest, or going to a psychiatrist and having oneself analysed, or engaging in one or another form of tedious, furtive and repressive discussion of the shadow. In the Pythagorean teaching, the shadow cannot understand itself. The shadow is void of the very possibility of self-knowledge. Real understanding can come solely through the light of self-awareness, which is inherent in every human being. The light of understanding can dispel the shadow of the personality only when, in lunar consciousness, a fruitful connection is made — metaphorically withdrawing to Metapontum where Pythagoras passed away, some say around the age of one hundred. Having built a bridge in personal consciousness towards the latent potential Self, one sees that in this larger selfhood there are no differences between oneself and every other human being and also the inner light and essence of anything and everything. The same luminous essence is to be found in a piece of paper, a table, a stone, in each single atom in space, in every animal form, in each vegetable and mineral, and the same is also to be found in every constituent of that vast and complex universe that we call the human body. The same is also to be found in each thought-form entering into and leaving the human mind through its affinity with appropriate centres of excitation in the brain, or when self-consciously drawn from an abundant cosmic storehouse.

All who want to come closer to the spirit of the *Golden Verses* must prepare and purge themselves as Pythagoras taught, thereby coming to be known as a trainer of souls. When human beings seek to learn, in the privacy and solitude of their own solemn undertaking, the serious business of truly elevating a human life, they must begin to ask questions about themselves: "Who am I?" "What am I?" "Why did I do this then?" "Do I always say what I intend?" "Did I think before I acted this morning, and what do I now think I am supposed to do tomorrow, next week, next year?" It is significant that the only phrase occurring twice in the *Golden Verses* is *Think before you act.* It is precisely because human beings with the best intentions in the world, with access to the profoundest ideas and sharing the noblest of feelings, are not able to deliver themselves in public life with the dignity of divine Monads, that they need to give themselves a chance, by making time within the space of every day for looking back in review. By continually reflecting the standpoint of the immortal Self, they will surely come to understand others and increase their real confidence through recognizing what is good in themselves; this in turn gives the courage to notice what in themselves is left-handed and must be discarded.

It is well known, though little understood, that in the Pythagorean School the psychological disciplines were joined to the study of mathematics. If one really wanted to understand this, one would be well advised to meditate deeply upon the Pythagorean Triad and the Tetrad. When one truly does so, one will find that the mystery deepens further, because what is esoteric and what is exoteric are relative. What is hidden to one is not unknown to another. What is hidden at one time is not inaccessible at another time. Unfortunately, many people are victims of an Aristotelian-Baconian view of knowledge where thoughts are seen as bits of information transmitted from the outside and impressed upon the brain, itself misconstrued as a kind of *tabula rasa*. In contemporary culture many people erroneously believe that true knowledge has got to do with the information revolution, and hence all that is needed is to find proper ways of giving access to information to each and all. In the School of Pythagoras, if people sought to know the Mysteries, they were fairly and squarely told what were the rules that must be respected. These time-honoured rules have always been observed. Great Teachers make fresh applications of these enduring rules according to the exigencies of the age in strict obedience to the Fraternity on behalf of which they act, and of which they are faithful members.

In teaching the Divine Wisdom relevant to his time, Pythagoras, the great Master, followed very strict rules. In one mythic version the story is told of how this was done. If individuals sought admission into the School, having already found inspiration in daily life from the ethical teaching of the *Golden Verses*, then they were invited to put themselves through a preliminary set of freely chosen and strictly administered tests. One of these required that the candidate be conducted to some secluded place and left with bread and water. He was requested to remain there for a night and to think intently upon a single symbol such as the triangle. Having prepared properly and taken whatever steps were needed to gain calmness, the candidate then set down ideas on the subject in relation to the whole of life. The following morning, the candidate was invited to the assembly of those who had already passed through these stages and asked by Pythagoras, who presided, to convey his observations to the entire group. A common practice during those days was that various members of the assembly were instructed to make it difficult for the candidate to state what he had to say by ridiculing his ideas. Naturally, a new candidate was liable to be nervous though the assembly was really on his side, yet nonetheless no concessions were made to his limitations, ambiguities and mixtures of motive. This was for his own good. Unless one could maintain one's composure under these circumstances, it was clear that life in the School would prove too much for a candidate who was unduly sensitive to criticism. Something of this ancient tradition still persists, for example in Holland and Germany at the time of the defence of dissertations, although without the compassionate purport of the trial the ceremony becomes censorious and even absurd.

What was crucial for Pythagoras was one's authenticity of self-knowledge in applying the holy and sacred teaching of the divine Triad for the sake of other men. The Triad itself could not be comprehended except in relation to the Point. The Point could not be grasped except as a One in relation to the Duad. The Monad and the Duad could not be understood completely unless they were also seen in terms of the Triad. And so the number series proceeds. Underlying it is the difficult problem which has to do with form, the meaning of the Pythagorean Square. If all of these are to be put together, something is involved which is rather like squaring the circle, securing the elixir of life, the key to the Mysteries of life and death. Pythagoras taught that unless the Mysteries are found within oneself, they cannot speak to one. All must make their own experiments with truth. They must make their own exercises in the calming of the passions, the controlling of the mind, the concentration of the thinking principle, and above all, the purgation and purification of their motives, intentions, feelings, likes and dislikes. This must be done for the sake of fusing the whole of one's being into an overriding thought-feeling, one keynote vibration which becomes a sacred Verbum, moving and animating the entire manifested self. All human beings have a unique and privileged access to the Verbum within the sanctuary of their own consciousness in deep sleep, in daily meditation, in waking life, in golden moments, but above all, when they begin to enter into a current of continuous thought and meditation upon the holiest of all subjects, which has to do with the *fons et origo* of all living things and beings. When they do this, then they will begin to understand the Tetraktys or sacred Quaternion, the Number of numbers.

Intuitive individuals will come to see that all these numbers point towards five, the Pythagorean Pentagon, and six, which was used later in the Kabbalah but for Pythagoras was a six-pointed star, where there was an eagle at the top and a bull and a lion below the face of a man. They will also begin to sense something about the significance of seven as the basic principle of division of not only colours and sounds, but of all manifestation. The seven in turn cannot be understood without the eight and Pythagoras taught how harmony may be produced when tuning the high and the low notes in the octave, thereby laying the basis for many of the theories and teachings that have come down through musical traditions. What he illustrated in music could also be applied to medicine, which means we cannot leave out the number nine. Nine has great meaning as three sets of three, but it also spells the ending of all things — incompletion. The wise take this into account in advance, thereby preserving the inviolable image of what since Pythagoras has come to be called the perfect number — ten — without seeking for its exact visual replica on earth. What is hidden in the Triad has been glimpsed by great architects, sculptors and craftsmen. The Chinese, when creating vases, abstained from making them perfectly symmetrical. Contemporary architects like Jacobsen, after conceiving a fine building, do not care to come to the opening ceremony, as they are absorbed in the designing of the next. Truly creative minds have known that there is a joy in creativity which is constricted and cancelled by attachment to results. The criteria of the world which accommodate the concerns of the mediocre also act as a brake upon the ascent to those levels of excellence which are relevant to all cultures. In the Pythagorean tradition, a proper answer to any question about the Mysteries must throw one back upon oneself so that each will do his own meditation and reflection upon the Tetrad as well as the Tetraktys.

The vital essence of the Pythagorean teaching was to encourage the emergence of whole men and women. They cannot be manufactured, but must truly create themselves. Great Teachers assist in the self-production of whole human beings by making a holistic teaching come alive. Pythagoras was an originator of true science, religion and philosophy in the Near Eastern cycle which he initiated. The teaching of Pythagoras was also that of the Buddha and later on of Shankara. Two thousand five hundred years ago the Buddha taught his disciples first to become *shravakas*, listeners. When they had spent a sufficient time in listening and learning, as in the earlier Hindu tradition with its emphasis upon *brahmacharya*, a period of probation, then they could become *shramanas,* men of action. We find this also in the Pythagorean tradition, where neophytes are *acousticoi*, those who listen. This has reference not to something mechanical or rigid and therefore false, but to a balanced training in the art of perfecting through wisdom the conservation of energy. The purification of thought, the calming and harmonizing of feelings, was undertaken for the sake of the appropriate manifestation of the Inner Self through proper speech and fitting conduct.

Pythagoras taught a threefold division of humankind and a threefold division of desire. All men may be compared to people who attend a festival. There are those who are motivated by the love of gain and who go to buy and sell. There are those motivated by the love of honour and they go to compete with and emulate each other in attaining standards of excellence. Then there are those who are concerned with neither gain nor glory because they have either worn out these toys, or thought through these illusions, or they are born with a natural indifference to them. Such are wholly concerned with the love of wisdom. Lovers of wisdom may be compared to those who at festivals are like spectators, not participating, but at the same time not making external judgements, not buying and selling, not comparing and contrasting, but merely learning what is common to all men, learning something about the noble art of living. They do not do what is unnecessary. They try to find out what is intimated behind the forms in the vaster human drama in which all the world is a stage and men and women merely players. The play is the thing. Quiet attention is the beginning of the way to wisdom in the Pythagorean tradition.

Reincarnation, the philosophy of palingenesis, is also fundamental. Every human being has been involved as a spectator in a variety of spectacles, has played a multiple diversity of roles. In this perspective, all learning is recollection, and much of what is seen is the restoration of soul-memory. What people think is new is mostly a recollection from where and whence they know naught, but which nonetheless acts as a divine prompting within them and sometimes saves them, in times of trouble and of trial, from making mistakes which would propel them further back than when they made them before, because by now they should have learnt something. The School which Pythagoras founded was one in which every kind of learning could be pursued, not for the purpose of integrating the isms and the sects of the time, but rather for coming down, from above below, so as to be able to see the synthesizing principles in *theoria* and *praxis*, contemplation and conduct.

After the passing of Pythagoras, the pupils of his School separated out. Schisms ensued between the so-called scientific people, who spent their time making claims, arguing and attacking each other, and those who initially espoused simple enthusiasms and were mocked by the others. The latter were left solely with their disarming trust, faith and devotion, which helped to continue the transmission of the tradition. All of this was known in advance by that wise Promethean called Pythagoras. He wanted separation and self-selection to take place not only amongst the many who were influenced, but equally amongst the few who were experiencing the rigours of training, those who had the moral fibre to endure the extremely difficult ascent to wisdom. The claim that the path is easy is the facile excuse of those who do not truly intend to make the ascent, because they have failed many times before and are inwardly so terrified of failure before they start that they would rather not risk even the first test.

There is much protection in the time-tested moral codes of every true community of seekers. This is suggested in the proverbs and the folklore of all societies. Pythagoras taught that there must be an inward quiescence of the soul, a stilling of the mind in which the true receptivity of the heart can enable real learning to take place. A person concentrating whilst learning carpentry, or whilst training for athletics, is quiet. Individuals who concentrate whilst preparing and studying for anything are quiet. Could any less be required of a person who would study and persevere whilst seeking the divine science of the dialectic, as Hierocles called the Pythagorean teaching? The art of free ascent of the soul towards the upper realms, indicated in the concluding words of the *Golden Verses*, is portrayed as the unveiling of latent perceptions of realities that are hidden. Anyone who is in earnest must give Nature time to speak. It is only upon the serene surface of the unruffled mind that the visions gathered from the invisible may find true and proper representation.

In ancient India, classical Greece and in early America it was well understood that without veneration for forefathers, nothing worthwhile can happen to a human being, a group or a society. This tradition was partly preserved under the influence of the Theosophical Movement in the nineteenth century and the subsequent short-lived Platonic renaissance in a variety of fraternities and movements. Some are still doing well, but most other fraternities, which took Pythagorean rules and adapted them for the purpose of self-discipline, true friendship and self-respect, are not in the same position. Whilst many have closed down, there are others that have held on though they lost the original impulse. There are also those few which have remained, and unknown to the many, have tried to be true to the original impulse. In some cases the impulse goes back not merely to the time of Benjamin Franklin or to the original societies of Philadelphia started at the time of the signing of the Declaration, but even earlier. As Burke suggested, any generation which fails to show respect to its ancestors will deserve nothing of posterity. Those who show little respect to those who have gone before them — their parents, grandparents, teachers and their teachers' teachers — will be repudiated in turn by their children. The law of Karma does not discriminate between persons, societies and generations.

The question came up amongst the early Pythagoreans regarding the injunction to honour one's parents: What is one to do if one's parents are unworthy? The answer given at the time by wise Pythagoreans was: First ask yourself whether you really have paid sufficient homage to the immortal gods, to the heroes of all time, and to the earth's good geniuses. If you have done all of these, then you are entitled to ask whether you should show honour to your parents. But you will find, if you have observed all that is prior, that you will always find some reason to honour your parents whilst at the same time you do not have to blindly follow their ways. That is because, as the later *Golden Verses* stress, all people must think for themselves. Each must make up his own mind and choose his own way. This does not require any recommendation or advertisement in our time. It is part of our very constitution. It was also the deathbed utterance of the Buddha. This is the oldest teaching, and it is common sense. There is hardly a human being who does not know it.

Human beings forget. All selfishness is rooted, Pythagoras felt, in thoughtlessness. It is hardly ever the case, even in the age of inversion, that people deliberately intend consistent and systematic inversion of reverence to the immortal gods, even though they may not know what that means, or to the heroes, even though they may have demythologized them. They do not deliberately intend to flout the law taught by an Initiate a long time ago: "God is not mocked; as ye sow, so shall ye also reap." Every man knows all of these things. Why then did Schweitzer put so much emphasis upon reverence for life? He knew that if something is worth doing, it is worth emphasizing, because men think they know it but act as though they never did. Men forget, and therefore in the Pythagorean doctrine of *anamnesis,* as in the Platonic teaching, everything has to do with remembering and forgetting. All human souls, when they have drunk of Lethe's waters, have become identified with forms and come under the influence of the lower languages transmitted to them by the world through their relatives and those near to them. They forget, and as they forget, the babies that stare and smile and greet the world in mystic wonder, in a very short time, in the process of learning how to toddle, to stand straight and to move, become confused in noticing the scorn and the scepticism, the cynicism and the distrust, the self-hatred that is all around. And by the time children are ready for the precious time of puberty, they have received no inspiration or help in learning to handle the sheer joy of using *eros* under the control of a calm and cool head. They are completely at war with others and with themselves.

We live in the age of Zeus, wherein it is difficult to understand the greatness, let alone the inner meaning, of the Pythagorean invocation of Brihaspati, Jupiter or Zeus — he who knows and can show the genius of every living being. Honour and reverence involve something more than the ordinary understanding of these words. They require what Pythagoras teaches in the closing stanzas of the *Golden Verses*, which collectively were called *Hieros Logos,* the Sacred Discourse. Pythagoras taught that discrimination and discernment are needed. One must learn not just to make distinctions but to show discrimination, to recognize nuances, sub-tones, sub-colours, shades of meaning, to recognize the immense diversity of forms of life, but also to see the ordering and the structure under which they could be understood. It is necessary to recognize similars, notice opposites, identify counterfeits, cherish intimations, but above all, to see the continuity and the connection between all of these. Then it will be possible, when hearing opinions, to discriminate amongst them and to go for the good and the gold in all, even in the most foolish observations. One can learn and note down what is of value in anything and everything that one comes across. But if one comes across a lot that is not worth entering into a notebook, one can let it go and remain calm in the presence of its utterance. All of this points to a conception of manhood, a magnitude of self-possession which combines with compassion and love, magnanimity as well as prudence, and which is truly rare in any age, but wholly admirable in our epoch.

Pythagoras especially commends prudence, not cunning or what the world calls shrewdness, but the insight of wisdom in relation to the lunar realm, a region in which everything that begins will change and pass away. If one does not remember this, one cannot be prudent. To be imprudent is to be over-attached. Desires are of three kinds. There are those desires which when they first arise are ill-fitting, inauspicious, and will do no good from the very start. Often such desires are longings to do the impossible. If a person, before being able to walk the Sierras, wants to climb Everest next week, it is an ill-timed and inauspicious desire. A wise friend might urge him to go and find out what in fact he needs to know, that such learning could be very unpleasant. The second kind of desire is not inauspicious to start with, but makes sense, like the desire to finish something which one has begun, whatever it be, whether it has got to do with school or job or family. Yet herein lies the danger of an immense inflation of that desire so that it becomes an obsession. It may become a virulent, over-mastering force, so that the person who has it is a slave and no longer a free human being. This kind of desire is not wholly bad, but it has got to be trimmed. The vehemence has got to be taken out of it until it runs like cool waters, consonant with the ocean of life into which it must eventually empty itself. Thirdly, there are desires which, though not unfitting to start with and though not vehement, become inappropriate in expression. A person might have a legitimate desire and a sense of proportion about it but not know how to express it appropriately, and hence become the frequent victim of bad timing. Bad timing is like bad faith, betokening a lack of total commitment and engagement in one's own project, to use Sartrean language. One is never quite there when needed but is always just that bit ill-timed. After a point one gathers around oneself elemental forces that become an ill-omened angel of misfortune.

When Pythagoras spoke of prudence and magnanimity, he gave a critical test. One is becoming a man when there is an increase in one's magnanimity. This teaching was so telling that even after Plato, with the decline of the Academy, Aristotle thought it fit to base his conception of the ideal man upon the quality of magnanimity. Every human being has access to magnanimity, but it cannot be secured instantly if one is mean, niggardly, fearful, selfish or contemptuous. Magnanimity is only released in the mind by large ideas and great visions. In the heart it is released only by a tremendous compassion for the sick and the suffering.

Pythagoras knew what was then to be judiciously chosen as a

foundation stone for the culture of the future. At present when modern culture is nearly dying and giving a great howl whilst doing so — but barely concealing a rather pathetic whimper — and another culture has already begun to come into birth as the invisible dips into the visible, the Pythagorean teaching cannot now mean a mere return to the forms once given in Magna Graecia. It must be seen and meditated upon as the seed of self-regenerating institutions and the culture and etiquette of the soul. When the soul becomes established firmly like a statue motionless in mind, whilst at the same time entertaining the vast universe of thoughts, the whole is fixed immovably in contemplation, showing beauty of soul, beauty of mind, beauty of heart, beauty in every direction and every dimension. It thereby makes it possible for more and more human beings, with their imperfections, to come out of the multitudes for the sake of all and for the sake of self-transformation and self-actualization, culminating in self-transcendence.

Preparations are crucial for the Pythagorean school of the future. Anyone who studies the *Golden Verses* of Pythagoras, in any translation or edition, and seeks by reflecting upon them to draw some inspiration, can release a vital energy in inward consciousness which is causal in relation to the external realm of effects. Those who do this could constitute themselves as beings who come closer in spirit, thought and feeling to the inmost, ever-unmanifest Presence. In the days of Pythagoras many people knew that they knew him not. No great Teacher ever incarnates or manifests except in proper conditions, and these are always hidden and always involve a few. When necessary he will manifest any relevant part of himself. Pythagoras spent a long time — twenty-two years — studying the Egyptian Mysteries, taking a projection of himself and letting it share all the ailments of the age. When he was ready to begin his work, he allowed people to see veiled appearances and partial expressions. His unmanifest and invisible Self, by its very nature (for those who apprehend the Golden Egg), can never be seen except by the light of the eye when the golden thread which is in every human being has been extended. This requires *Buddhi*. One who reads the *Golden Verses* in this reverential spirit can come closer to the Divine Being who was their wise author and gain inspiration which would be invaluable in times of trouble.

HIEROS LOGOS

THE GOLDEN VERSES OF PYTHAGORAS

In the first place revere the Immortal Gods, as they are established and ordained by the Law.

Reverence the Oath. In the next place revere the Heroes who are full of goodness and light.

Honour likewise the Terrestrial Daimons by rendering them the worship lawfully due to them.

Honour likewise thy father and thy mother, and thy nearest relations.

Of all the rest of mankind, make him thy friend who distinguishes himself by his virtue. Always give ear to his mild exhortations, and take example from his virtuous and useful actions. Refrain, as far as you can, from spurning thy friend for a slight fault, for power surrounds necessity.

Know that all these things are as I have told thee.

Accustom thyself to surmount and vanquish these passions: First, gluttony, sloth, lust and anger. Never commit any shameful actions, neither with others nor in private with thyself.

Above all things, respect thyself.

In the next place, observe Justice in thy actions and in thy words; and accustom not thyself to behave thyself in anything without rule and without reason.

Always make this reflection, that it is ordained by Destiny for all men to die; and that the goods of fortune are uncertain. As they may be acquired, they may likewise be lost.

Concerning all the calamities that men suffer by Divine Fortune, support with patience thy lot, be what it will, and never repine at it, but endeavour what thou canst to remedy it, and consider that Fate does not send the greatest portion of these misfortunes to good men.

There are amongst men several sorts of reasonings, good and bad. Admire them not too easily and reject them not either, but if any falsehoods be advanced, give way with mildness and arm thyself with patience.

Observe well, on every occasion, what I am going to tell thee: Let no man either by his words, or by his actions, ever seduce thee, nor entice thee to say or to do what is not profitable for thee.

Consult and deliberate before thou act, that thou may'st not commit foolish actions, for it is the part of a miserable man to speak and to act without reflection.

But do that which will not afflict thee afterwards, nor oblige thee to repentance.

Never do anything which thou dost not understand; but learn all thou oughtest to know, and by that means thou wilt lead a very pleasant life.

In no wise neglect the health of thy body; but give it food and drink in due measure, and also the exercise of which it has need. By measure, I mean what will not incommode thee.

Accustom thyself to a way of living that is neat and decent, without luxury. Avoid all things that will occasion envy, and be not expensive out of season, like one who knows not what is decent and honourable.

Be neither covetous nor niggardly. A due measure is excellent in these things!

Do only the things that cannot hurt thee, and deliberate before thou doest them.

Never suffer sleep to close thy eyelids after thy going to bed, till thou hast thrice reviewed all thy actions of the day: Wherein have I done amiss? What have I done? What have I omitted that I ought to have done?

If in this examination thou find that thou hast done amiss, reprimand thyself severely for it; and if thou hast done any good, rejoice.

Practise thoroughly all these things; meditate on them well; thou oughtest to love them with all thy heart. It is they that will put thee in the way of Divine Virtue.

I swear it by Him who has transmitted into our souls the Sacred Tetraktys, the Source of Nature, whose course is eternal.

Never set thy hand to the work, till thou hast first prayed the Gods to accomplish what thou art going to begin.

When thou hast made this habit familiar to thee, thou wilt know the constitution of the Immortal Gods and of men; even how far the different Beings extend, and what contains and binds them together.

Thou shalt likewise know, in accord with Cosmic Order, that the nature of this Universe is in all things alike, so that thou shalt not hope what thou oughtest not to hope; and nothing in this world shall be hid from thee.

Thou wilt likewise know that men draw upon themselves their own misfortunes, voluntarily and of their own free choice.

Wretches that they are! They neither see nor understand that their good is near them. There are very few of them who know how to deliver themselves out of their misfortunes.

Such is the Fate that blinds mankind and takes away his senses. Like huge cylinders, they roll to and fro, always oppressed by ills without number; for fatal contention, which is innate in them, pursues them everywhere, tosses them up and down, nor do they perceive it.

Instead of provoking and stirring it up, they ought by yielding to avert it.

Great Jupiter, Father of men, you would deliver them all from the evils that oppress them, if you would show them what is the Daimon of whom they make use.

But take courage, the race of men is divine. Sacred Nature reveals to them the most hidden Mysteries.

If she impart to thee her secrets, thou wilt easily perform all the things which I have ordained thee, and healing thy soul, thou wilt deliver it from all these evils, from all these afflictions.

Abstain thou from all that we have forbidden in the Purifications; and in the Deliverance of the Soul make a just distinction of them; examine all things well, leaving thyself always to be guided and directed by the understanding that comes from above, and that ought to hold the reins.

And when, after having divested thyself of thy mortal body, thou arrivest in the most pure Aether, thou shalt be a God, immortal, incorruptible, and death shall have no more dominion over thee.

THE COMMENTARY OF HIEROCLES
ON THE GOLDEN VERSES

Philosophy is the purification and perfection of human nature: its purification, because it delivers it from the temerity and folly that proceed from matter and because it disengages its affections from the mortal body; and its perfection, because it makes it recover its original felicity by restoring it to the likeness of God.

Now virtue and truth alone can operate these two things: virtue, by driving away the excess of the passions, and truth, by dispelling the darkness of error and by returning the divine form to such as are disposed to receive it.

For this science, therefore, which ought to render us pure and perfect, it is good to have short and certain rules as so many aphorisms of the art, that by their means we may arrive methodically and in due order at happiness, which is our only end.

Amongst all the rules that contain a summary of Philosophy, the Verses of Pythagoras, called the *Golden Verses,* justly hold the first rank, for they contain the general precepts of all Philosophy regarding the *active* as well as the *contemplative* life. By their means everyone may acquire truth and virtue, render himself pure, and happily attain to the Divine Resemblance. As is said in the *Timaeus* of Plato (whom we ought to regard as a very exact master of the doctrine of Pythagoras), after having regained his health and recovered his integrity and his perfection, he may see himself again in his primitive state of innocence and of light.

Pythagoras begins by the precepts of *active* virtue. Before all things, we ought to dissipate and drive away the folly and the laziness that are in us and then apply ourselves to the knowledge of divine things. As an eye that is diseased and not yet healed cannot behold a dazzling and resplendent light, in like manner a soul that is still destitute of virtue cannot fix its view on the beauty and the splendour of truth. Nor is it lawful for impurity to touch the things that are pure.

Practical Philosophy is the mother of virtue, and contemplative virtue is the mother of truth, as we are taught by these very Verses of Pythagoras, where Practical Philosophy is called *Human Virtue* and where the Contemplative is celebrated under the name of *Divine Virtue*. After having finished the precepts of civil virtue by these words: *Practise thoroughly all these things; meditate on them well; thou oughtest to love them with all thy heart*, he continues, *it is they that will put thee in the way of Divine Virtue*.

We must, therefore, first be men and afterwards become God. The Civil Virtues make the man and the sciences lead to Divine Virtue which makes the *God*. Now according to the rules of the Order, little things must precede the greater if we would make any progress, and this is the reason why in these Verses of Pythagoras the precepts of virtue are the first to teach us that the practice of virtues, which is so necessary in this life, is the way whereby we ought to advance and rise even to the Divine Image. And the order and design proposed in these Verses is to give to those that read them the true character of Philosophy before they are initiated in the other sciences.

They are called *Golden Verses* to signify to us that they are the most excellent and most divine of any of this kind. In like manner we call the *Golden Age* the age that produced the greatest men, and describe the difference of the manners of the several ages by the analogical qualities of metals. Gold being the purest of all metals and free from all the dross found in the other metals that are inferior to it — as silver, iron and brass — is therefore the most excellent, being the only metal that never breeds any rust, whereas the others grow rusty in proportion to the quantity of dross they have mixed in them. Rust, therefore, being the figure and emblem of vice, it was but reasonable that the age in which sanctity and purity reigned and which was exempt from all corruption of manners should be called the *Age of Gold*. Thus these Verses, being every way sovereignly good, have justly deserved the appellation of *Verses Golden and Divine*. We find not in them, as in all other poems, one good verse and another that is not so, but they are all perfectly good, they all equally represent purity of manners, lead to the likeness with God, and discover the most perfect aim of the Pythagorean Philosophy, as will evidently appear by the explanation we are going to give of each Verse in particular.

Let us then begin with the first.

VERSE 1

In the first place revere the Immortal Gods, as they are
established and ordained by the Law.

Since the piety that relates to the Divine Cause is the chief and
the guide of all the virtues, the precept concerning that piety is
with good reason placed at the head of all the laws prescribed by
these Verses. We ought to honour the Gods of this Universe
according to the Order in which they are established, and which
the Eternal Law that produced them distributed them with their
Essences, placing some of them in the first sphere of Heaven,
others in the second, others in the third and so on, till all the
celestial globes were filled up. To acknowledge and honour them
according to the order and station in which they were placed by
their Creator and Father is to obey the Divine Law and to render
them truly all honour due to them. Nor ought we to extol their
dignity above measure any more than to entertain diminishing
thoughts about them, but we should take them for what they are,
give them the rank they have received, and refer all the honour we
render them to God alone who created them and who may
properly be called the God of Gods, the most high and most good
God. The only way we have to discover and comprehend the
majesty of this excellent Being who created the world is to be
fully convinced that He is the Cause of the Gods and the Creator
of the Rational and Immutable Substances. These are the
Substances, these the Gods we here call *Immortal Gods.* They have
always the same opinion and the same thoughts of God who
created them because they are always intent upon this Supreme
God and united with Him. They have received from Him immutably
and indivisibly the being and the well-being too, inasmuch as they
are the unchangeable and incorruptible Images of the Cause that
created them. For it is worthy of God to have produced such
Images of Himself as were not capable of change nor of corrupting
themselves by their inclinations to ill as are the souls of men, who
are the last of all Intelligent Substances, whilst those that are called
Immortal Gods are the first.

And it is to distinguish them from the souls of men that we here
call them *Immortal Gods,* because they never die and never forget,
one single moment, either their own Essence or the goodness of

the Father who created them. Consider the passions and alterations to which the soul of man is subject: sometimes it remembers its God and the dignity in which it was created, and sometimes it entirely forgets both the one and the other. For this reason the souls of men may justly be called *Mortal Gods,* as dying sometimes to the Divine Life by their going astray from God and sometimes recovering it again by their return to Him, living thus in this last sense a life divine and in the other dying as much as it is possible for an Immortal Essence to participate in death — not by ceasing to be, but by being deprived of well-being. For the death of a Reasonable Essence is ignorance and impiety, which drag after them disorders and revolt of the passions, and the ignorance of good necessarily plunges us into the slavery of ill — a slavery whence it is impossible to be redeemed except by returning to knowledge and to God, which is done by recollection and the faculty of reminiscence.

Now between these *Immortal* and *Mortal Gods,* as I have called them, it is necessary that there should be an Essence superior to man and inferior to God, to be, as it were, a medium and a link to chain the two extremes to one another, to the end that the whole Intelligent Essence might be bound and united together.

This middle Essence, *the Angels,* is never altogether ignorant of God, yet has not always an equally immutable and permanent knowledge of Him, but sometimes a greater, sometimes a less. By this state of knowledge which never absolutely ceases, it is superior to the nature of man, and by this state of knowledge which is not always the same but lessens or increases, it is inferior to the nature of God. It has not raised itself up above the condition of man by its proficiency and improvement in knowledge, and it is not become inferior to the *Gods.* Nor has it been placed in this middle rank by reason of the diminution of the same knowledge, but it is by its nature a mean, a middle Being. For God who created all things established these three Beings, first, second and third, different from one another by their nature; nor can they ever displace themselves nor confound themselves one with another either by vice or by virtue. But being eternal by their nature, they differ according to the rank that has been given them. They were placed in this order because of the causes that produced them. It is Order that contains the three degrees of perfect wisdom, the first, the second and the third. Wisdom is wisdom only because it produces its works in order and perfection, inasmuch as wisdom, order and perfection are always found together and never separate from one another. In like manner, in this Universe the Beings produced by the first thought of God ought to be the first in the world, those that are produced by the second, the second or middle, and those that resemble the end of the thoughts, the last of all rational Beings. This whole reasonable Order with an incorruptible body is the entire and perfect image of God who created it. The Beings that hold the first rank in this world are the pure Image of what is most excellent in God; those that hold the middle rank are the middle Image of what is middling in God; and those that hold the third, the last rank among the rational Beings, are the last Image of what is last in the Divinity. The first of these Orders is here called the *Immortal Gods,* the second *Heroes who are full of goodness and light,* and the third *Terrestrial Daimons,* as we shall see hereafter.

Let us now return to what we were saying. What is the Law? What is the Order that is conformable to it? And lastly, what is the honour rendered in regard to this Order and to this Law? The Law is the Intelligence that has created all things; it is the Divine Intelligence by which all has been produced from all eternity and which likewise preserves it eternally.

The Order conformable to this Law is the rank which God, the Father and Creator of all things, gave the *Immortal Gods* when He created them, and that appoints some of them to be first and others second. For though they are the first in all this Intelligent Order and have received whatever is most excellent, they are different, nevertheless, amongst themselves — some are more, some less, divine than the others. A mark of their superiority or inferiority with regard to the others is the rank and order of the Celestial Spheres, which were distributed amongst them according to their Essence, power and virtue, inasmuch as the Law relates only to their Essence and the Order is only the rank that was given to them suitable to their dignity. For neither were they created fortuitously nor separated and placed by chance, but they were created and placed with order, as different parts and different members of one single *Whole,* which is Heaven. They preserved their connection in their separation and in their union according to their kind so that no change, no displacing, can be imagined in their situation without the ruin of the world. This can never happen so long as the First Cause that produced them continues immutable and firm in His decrees and has a power equal to His Being, and shall possess a goodness not acquired but inherent and essential to Him, and, for the love of Himself, shall promote the good and happiness of all creatures. For no other reasonable cause of the creation of things can be alleged than the essential goodness of God. It is God who is all good by His nature, and what is good is never susceptible of the least envy. All the other causes that are assigned for the creation of the Universe, except the sole goodness of God, savour more of the necessities and of the wants of men than of the independence of an Almighty God.

Now God, being all good by nature, produced first the Beings that most resemble Himself, secondly, those of a middle likeness to Himself, and thirdly, those who of all the Beings that resemble Him, participate the least in His Divine Image.

The Order was regulated according to the Essence in all these created Beings, inasmuch as what is more perfect is preferred to the less perfect, not only in all the kinds, but likewise in the different species of each kind. For it was neither by chance nor by change of choice or will that all things received their place and their rank. But having been created different by the Law that produced them, they have the rank that best agrees with the dignity of their nature. Therefore this precept, *Revere them as they are established and ordained by the Law,* ought to be understood not only of the *Immortal Gods,* but also of the *Heroes, the Angels,* and of the souls of men. For under each genus there is an infinite number of species, placed and disposed according to their greater or lesser dignity. Thus you see what is the nature and what the order or rank of Intelligent Beings.

What is, then, the Law, and what the reverence that is the consequence of it? Let us repeat it once again: The Law is the immutable power of God according to which He created the Divine Essences and ranked and placed them from all eternity in an Order they can never change. And the reverence conformable to this Law is the knowledge of the nature of these Beings which we revere, and the likeness which as much as possible we labour to have with them. For whatever we love, we imitate as much as we can, and the reverence we offer to Him who has no need of anything consists in receiving the good things He offers us. For thou dost not honour God by giving Him anything but by rendering thyself worthy to receive from Him, and as the Pythagoreans say: *Thou wilt honour God perfectly if thou behave thyself so that thy soul may become His Image.* Every man who honours God by gifts, as a Being that has need of them, falls unthinkingly into the error of believing himself greater and more powerful than God. Even the magnificence of gifts and offerings is no honour to God unless a heart truly penitent offers them. For the gifts and the victims of fools are only fuel for the flames, and their offerings but a bait for the sacrilegious. But a mind truly penitent and sufficiently strengthened and confirmed in love unites itself to God, and it is of necessity that the like should have a tendency to its like. For this reason it is said that the wise man is the only sacrificer, that he alone is the friend of God and knows how to pray. He alone knows how to revere who never confounds the dignity of those he honours. He offers himself first as a pure sacrifice, renders his soul the Image of God, and prepares his mind as a temple worthy to receive the Divine Light. . . .

VERSE 2

Reverence the Oath.

We have shown that the Law is the power of God by which He operates and brings all things to pass immutably and from all eternity. And here, in consequence of this Law, we say that the *Oath* is the cause that preserves all things in the same state, since they are made firm and stable by the faith of the *Oath* and preserve thereby the Order established by the Law, so that the unchangeable disposition of the created Beings is only the effect of the Law that produced them and of the *Oath* that maintains and secures them. That all created Beings continue as they were disposed and set in order by the Law is the chief work and the first effect of the *Divine Oath*, which is above all and always observed by those whose thoughts are continually bent on God, but is often violated by those that think not always of Him and sometimes forget Him. And, indeed, they violate the *Oath* proportionately as they withdraw and go astray from God, and keep it proportionately as they return again to Him.

For by the *Oath* in this place is meant only the observance of the Divine Laws and the bond by which all created Beings are linked to God the Creator to the end that they may know Him. Some amongst those creatures which are always united to Him in *reverence of the Oath* sometimes apostasize from Him, thereby rendering themselves impious violators of this *Oath*, not only by transgressing the order of the Divine Law, but also by breaking the faith of the *Divine Oath*. And such is the *Oath* that we may call it *innate* and *essential* to Intelligent Beings who must keep themselves always only united to God, their Father and Creator, and never transgress the Laws that He has established.

But the oath to which men have recourse in the affairs of the civil life is the shadow, as it were, the copy of this original *Oath*, and it leads directly to truth those that make use of it as they ought. By dissipating the ambiguity and uncertainty of the designs of men, it renders them plain and certain. It fixes and forces them to continue such as they are declared to be, either in words or in actions, by discovering on one hand the truth of what is already done, and by exacting and securing on the other what is yet to be done. Thus you see the great reason why oaths ought, above all things, to be religiously observed.

The first, which precedes by its Essence, claims our respect and observance as the pledge of eternity. The human oath, which is a certain help to us in the affairs of life, ought to be respected as the image of the first and as that which, next to the *Divine Oath,* is the safest depository of certainty and of truth, and adorns and enriches with very excellent morals all who have learnt to respect it.

Now the respect due to an oath is the most faithful and most inviolable observance possible of what we have sworn to; this observance is the virtue that associates and unites with the firm stability and truth of the Divine Habitude those that respect and keep their oaths freely and voluntarily.

The unspeakable sanctity of the first *Oath* may be recovered by a sincere conversion to God, when by the purifying virtues we heal the breach of this *Divine Oath*. But the sacredness and fidelity of a human oath is preserved by politic virtues, for they alone who profess those virtues can be faithful in the oaths of the civil life. And vice, the father of infidelity and of perjury, tramples oaths under foot through the instability and inconstancy of manners.

How can the covetous man be faithful when he is to receive or pay money? How can the intemperate and the cowardly religiously observe their oaths? Will not either of them, whenever they believe it will be to their advantage, cast off all respect for what they have sworn to perform, and renounce eternal happiness for the enjoyment of frail and temporal goods? Only those who never deviate from the paths of virtue are capable of preserving the respect that the majesty of an oath requires.

Now the most certain way to preserve this respect inviolably is not to make use of oaths frequently nor rashly, nor by chance, nor for things of little concern, nor as an ornament of discourse, nor the more to ascertain what you say, but to reserve it for things necessary and honourable and for those occasions only where there appears to be no other way of safety for you in your affairs than by the truth of an oath. And the only way to convince all that hear us of the truth of what we affirm is so to behave ourselves that our manners may agree with our oaths, and not give our neighbour any cause to suspect that we are capable of preferring any temporal advantage whatsoever to truth, whether we have or have not obligated ourselves by an oath.

This precept, *Reverence the Oath*, commands us not only to be true and faithful to our oaths, but likewise to abstain from swearing, for not to swear too frequently is the surest way to swear true always. . . .

Now the faithful observance of an oath agrees perfectly with the honour the first Verse commands us to pay to the *Gods*, for it is the inseparable companion of piety. Thus an *Oath* is the guard and security of the Divine Law for the order and disposition of the Universe.

Honour, then, this Law by being obedient to what it commands, and respect an oath by not making use of it at every turn, that thou mayest accustom thyself to swear true by avoiding the habit of swearing. For the truth of an oath is no small part of piety. But we have said enough concerning the first Beings, concerning the Divine Law which is the Author of order, and concerning the *Oath* which is the consequence of the Law.

Now because next to the *Immortal Gods* we ought to honour the Beings we call angelical, the author of these Verses goes on: —

In the next place revere the Heroes who are full of goodness and light.

These are the middle sort of the Intelligent Essences, and holding the next place after the *Immortal Gods,* they precede human nature and join the last Beings to the first.

Since, therefore, they hold the second place, we ought to render them the second honour by understanding likewise in regard to them these words of the first precept: *Revere them, as they are established and ordained by the Law.* All the virtue and force of this honour consists in truly knowing the Essence of those we honour, for how can we address ourselves in due manner to them we know not, and how shall we offer presents to them of whose dignity we are ignorant? The first, therefore, and only true reverence in regard even to those *Heroes full of goodness and light,* is the knowledge of their Essence and of their rank, along with a precise and true discernment of their employments and of the perfection they contribute on their part to this Universe in consequence of the rank they hold. We ought in all things to proportion the reverence we pay them to their Essence; and this proportion can proceed from nothing but from the knowledge we have of their divinity. When we once know the nature and the rank of each Being, then, and then only, we shall be able to render them the reverence they deserve and that which the Law commands us to render them.

We are to revere no nature inferior to human nature, but we are chiefly to revere the Beings that are superior to us by their Essence and those who, having been our equals, have distinguished and raised themselves above us by the pre-eminence of their virtues.

Of all the Beings superior to us by their Essence, the first and most excellent is God, who created all things, and it is He, too, who ought to be revered above all, without any comparison or competition. They who are next to Him and by Him the first in the world, whose thoughts are continually bent on Him and who express and represent faithfully in themselves all the good which the Cause has created and whom the first Verse calls *Immortal Gods,* I say, ought to receive the first honours after God. The second and middle honours are due to the middle Essence, that is to say, to those who hold the second rank and are here called *Heroes full of goodness and light,* who think without ceasing on their Creator and who are all resplendent with the light that reflects from the felicity they enjoy in Him, though not always in the same manner nor without any change. Since they are united to God as middle Essences and have received the grace of being always turned towards Him (without it being in their power ever to depart or go astray from Him), they continue always in the presence of this First Being but with efforts that are not always equal. By the full and entire knowledge they have of themselves, they divide and reunite the unchangeable intimateness that the first Beings have with God, by making intimateness with these Beings the beginning of their Initiation. Therefore they are with reason called *Excellent Heroes,* the epithet that signifies *excellent* intimating to us by its root that they are full of goodness and understanding, and the word *Heroes,* coming from a word that signifies 'love', to show us that since they are full of love for God, their whole endeavours are to assist us in our passage through this terrestrial life to a life divine and to help us to become citizens of Heaven. They are likewise called *good Daimons,* as they are instructed and knowledgeable in the Divine Laws. Sometimes we give them the name of *Angels* because they declare and announce to us the rules that will assist us to live well here, and lead us to happiness hereafter. Sometimes, too, according to these three senses, we divide into three classes all those middle Spirits. Those that approach the nearest to the Celestial and Divine Essences we call *Angels;* those that are united to the Terrestrial Essences we call *Heroes;* and those that hold the middle place, equally distant from the two extremes, we call *Daimons,* as Plato frequently divides them.

Others give to this middle kind but one of these three names, calling them either *Angels* or *Daimons* or *Heroes,* for the reason we have already given; and thus the author of these Verses has called them *Heroes full of goodness and light,* for they are, with respect to the first kind, as light is to fire and as the father is to the son. Therefore they are celebrated, and with justice, too, as the Children of God, for they are not born of mortal race but are produced by their uniform and only Cause as light comes from the Essence of a luminous body (I mean a pure and clear light), after which it is easy to imagine a light full of shades and blended with darkness. And to this obscure and dim light analogically answers the third kind of Beings. I mean mankind, by reason of the proneness they have to vice and to oblivion, which makes them incapable of contemplating God always. They are inferior to the Beings that always think on Him because they cease sometimes to have Him in their thoughts, and this is their darkness. But they are superior to the Beings void of reason because they return sometimes to think on God, and are now and then restored to the Divine Knowledge when they join themselves to the Celestial Choirs by laying aside all carnal affections and disengaging themselves from the corruption of the body; and this is their light.

He who is favoured with this divine grace becomes worthy of our homage and respect by having adorned and raised up the equality of our nature by participation in what is most excellent.

Now every man that loves God ought likewise to love every Being that in any way resembles Him, whether it has possessed this likeness from all eternity or has acquired it in time, such as those men who have distinguished themselves by the pre-eminence of their virtues. The following Verse concerns this precept.

VERSE 4

Honour likewise the Terrestrial Daimons by rendering them the worship lawfully due to them.

The author of these Verses, speaking of the souls of men who are adorned with truth and with virtue, calls them *Daimons* because they are full of knowledge and light. Afterwards, in order to distinguish them from the *Daimons* that are such by nature and that hold the middle rank, as has been said already, he adds the epithet *Terrestrial,* to show that they can converse with man, inform and animate mortal bodies, and dwell upon the earth.

By calling them *Daimons* he distinguishes them from wicked

and impious men, who are ignorant and consequently far from being *Daimons*. And by adding the epithet *Terrestrial*, he distinguishes them from those that are always full of light and knowledge and who are not of a nature to live upon the earth nor to animate mortal bodies. For this name of *Terrestrial Daimon* is applicable only to him who, being man by nature, is become *Daimon* by habitude, by his union with and knowledge of the things relating to God. The third sort are called purely and properly *Terrestrial Daimons,* being the last of the reasonable Substances and entirely addicted to a terrestrial life. For the first sort is Celestial, and the second or middle sort is Aethereal. Thus, therefore, all men being *Terrestrial,* that is to say, holding the third and last rank amongst intelligent Substances and not being all of them *Daimons,* that is to say, endowed with knowledge and understanding, the author of these Verses was in the right to join these two names, *Terrestrial Daimons,* to signify wise and virtuous men, for all men are not wise and all the wise are not men. The *Heroes* and the *Immortal Gods* are by their nature much superior to men and likewise endowed with wisdom and virtue.

This Verse, therefore, commands us to respect and revere the men who have been admitted into the Celestial Orders and who may be considered as equal to the *Daimons,* to the *Angels* and to the *Heroes.* We are not to imagine that we are here advised to respect and honour any vile and contemptible sort of *Daimon* (as the common acceptation of the word *Terrestrial Daimon* might be apt to persuade us), for in a word, all the Beings that are inferior to human nature ought in no wise to be honoured by those who are touched with the love of God and have a sense of their own nobility and dignity. Nor are we to honour man himself next to the superior Beings unless he has become like them and has been received into the Divine Choir.

What is, then, the honour and respect we owe them? *To render them,* says this Verse, *the worship lawfully due to them.* And this worship consists only in obeying the precepts they have left us and in regarding those precepts as Laws that must not be violated; to take example by their way of living; and to walk in the paths they walked in, which envy could never hinder them from teaching us, and which they have transmitted to their successors with ten thousand toils and labours as the immortal inheritance of their

34

Fathers by consigning to us in their writings the elements of virtue and the maxims of truth. . . .

Honour likewise thy father and thy mother, and thy nearest relations.

The author, having in the foregoing precept commanded us to honour and revere good and virtuous men as divine Beings who enjoy eternal felicity, now exhorts us to honour our father and mother and those that are in any way related to us, upon the same necessity of kindred. For as the superior Beings, the Celestial, stand to us as parents, and the *Heroes* as relations — by means of the bond and union that has been and is between them and us from all eternity — so our fathers and our mothers and their relations next of blood, who for that reason ought to receive from us the first honours after our parents, are the same thing to us in this mortal life. How, then, shall we honour them? Shall we behave as they would have us, so as neither to think nor do anything but what will please them? In this way our zeal for virtue will degenerate into zeal for vice if our parents and relations happen to be wicked and vicious. But on the other hand, shall we neglect and condemn them because we know them to be vicious?

How, then, are we obedient to this Law? Can we by not honouring our parents, who are the Image of the *Gods,* or our relations, who represent to us the *Heroes,* can we, I say, be other than impious to those *Gods* and *Heroes* whom we agree that our parents and relations resemble? And will this virtue which we think we practise by disobeying our parents, because of their vices, produce a greater ill — impiety? And if, on the contrary, we obey them in everything, how can we do otherwise than depart from the practice of virtue and piety if it should happen that through the corruption of their manners they instructed us not to direct our sons in the paths of virtue and truth? For if whatever our parents commanded us were true and good, the honour we rendered them would perfectly agree with the honour and obedience we owe the *Gods.* But if the will of our parents is not always conformable to the Laws of God, what will they do who find themselves in this sort of contrariety between the two Laws, but is daily practised in other duties that on some occasions happen to be incompatible and contradictory to one another, and where one must necessarily be violated that the other may be observed? Of two good actions that offer themselves to us — the one whereof is barely good, the other better — we ought indispensably to prefer the better when we cannot acquit ourselves of both. It is a good action to be obedient to God, and it is good likewise to obey our parents. If what God and our parents require of us agree, so that in obeying both we tend to the same end, it is a great happiness for us, and this double duty is indispensable. But if the Law of God commands us one thing and our parents another, we ought, in this contradiction which cannot be reconciled, to obey God by disobeying our parents in the things only wherein they themselves are disobedient to the Divine Laws. For it is not possible that a man who would exactly observe the rules of virtue should ever agree with those that violate them.

In all other things we ought to honour our parents to the utmost of our power and without any limitation, by serving them ourselves and by supplying them abundantly with all our heart with the things they have need of. For it is most reasonable they should rely upon those to whom they gave life and learning.

But in regard to what we received not from them, the Law declares it free and exempts it from their power, commanding us to seek the true Father of it, to unite ourselves to Him, and to labour particularly to render ourselves conformable to His Image. Thus we shall be able to preserve both divine and human goods. And as we ought not to neglect our parents under a vain pretext of virtue, so neither ought we to fall by a blind and senseless obedience into the worst of all evils — impiety. But if they threaten to put us to death for our disobedience or to disinherit us, we ought not to be dismayed at their ill will, but think immediately on what it will fall. They threaten only what they made. But as to that part of us that is safe from their passion, that cannot suffer by their injustice and that comes not from them, we ought to preserve it free and subject to the will of God.

The true honour that virtue commands us to render to our parents is to spare nothing to serve them, neither in body nor in goods, but to be entirely subject to them in what concerns either of the two, for it is becoming and reasonable never to refuse them the service of our hands. On the contrary, the more this service is toilsome, mean and servile, the more ought we to delight in it and think ourselves honoured by it. Much less ought we to refuse to supply their wants and grudge to lessen their expense through a motive of avarice, but we ought rather to be lavish in furnishing them with all they have occasion for and to do it with a cheerful mind, thinking ourselves happy in serving them with our persons and estates. To practise these two things joyfully and with a free will is to fulfil the law of virtue and to satisfy the rights of nature.

Thus you see what is the honour due to our parents. What we owe to their relations, which is only the second honour, must be proportioned according to the degree of consanguinity, so that next to our parents we ought more or less to honour our relations according to the rank of affinity in which Nature has placed them.

VERSE 6

Of all the rest of mankind, make him thy friend who distinguishes himself by his virtue.

After the law that prescribes the first honour that is due to our first affinity, and after that which regulates the respect we owe our parents and their relations and which is a consequence of the first, follows immediately the law concerning the contracting of friendship. This law obliges us to choose for our friend, amongst those who are not of our family, one who is the most virtuous, and to bind ourselves to him by the communication of virtues, to the end that we may make the good man our friend for a good reason and not seek his friendship on any other account. This precept is entirely conformable to the advice that was given us concerning the good men who have departed this life.

For as we were told that we ought to honour and revere only those who are full of knowledge and light, so likewise we are told that we ought to contract friendship with none but men of probity and virtue. As to these we are allowed the liberty of choice; as to our parents and relations we are confined to obey the dictates of Nature, for a father or a brother naturally claims our respect. But as for others — I mean our friends — it is virtue only that makes them valuable, in the same manner that it gives merit to the dead. . . .

Always give ear to his mild exhortations, and take example from his virtuous and useful actions. Refrain, as far as you can, from spurning thy friend for a slight fault, for power surrounds necessity.

We are now going to show how we ought to behave towards our friends.

First, we must yield to and obey them when they exhort us to virtue and when they do anything for our advantage. It is for our mutual good that the Law of Friendship binds us together, to the end that our friends may assist us in the increase of virtue and that we may reciprocally assist them in their improvement therein. For as fellow travellers in the way that leads to a better life, we ought, for our common advantage, to impart to them the good things we may discover, perhaps better than they.

We ought mildly to submit to the good advices of our friends and to let them share with us in whatever we have that is good and profitable. As for riches, glory and all other frail and perishable goods, we ought never to differ in the least with our friends concerning them, for that would be to hate for a slight offence those who are our friends in matters of the highest moment.

Let us then bear with our friends in all things, as being bound to them with the strictest of all bonds — the sacred tie of friendship. There is but one thing which we are not to bear with in a friend, and that is when he falls into a corruption of virtue.

And we are in no wise to follow his example when he quits the ways of wisdom and takes up another course of life, for then we should suffer ourselves to be seduced and led astray by his example. But we ought to use our utmost endeavours to reclaim our friend and to bring him back into the good way.

If we cannot prevail with him to return, we must rest satisfied and not regard him as our enemy because of our former friendship, nor as our friend because of his depravity. For this reason only we ought to renounce and forsake him because he has become incapable on his part of assisting or forwarding us in the pursuit and improvement of virtue, for which cause alone we sought his friendship.

But let us take great care that this separation degenerate not into enmity, for though he first unlinked the chain, we are obliged to leave no means untried to reclaim him to his duty, without any rejoicing at the fall of our friend and without insulting him on account of his error. Rather be compassionate at his misfortune with tears and with sorrow, and pray for him and forget nothing that may bring him to repentance and procure his welfare. . . .

Let us then be fully persuaded that we ought to bear with our friends as much as necessity shows it is possible for us to do, and that on account of the relation of friendship we ought to endure what seemed to us insupportable. For we are not to imagine that courage and fortitude are never to be employed except when we are to resist the efforts of strength and violence. Whatever tends to the preserving or to the regaining of our friends requires and deserves more patience as being the injunctions of a Divine Necessity.

Now the Necessity of Reason is stronger and more prevalent with the wise than all exterior force. Whether, therefore, thou regard the necessity that arises from the several conjunctures and circumstances of affairs, or whether thou consider the Necessity of the will, thou wilt find this last, this free and independent Necessity, this voluntary result of Reason which is contained within the bounds of knowledge and is an emanation from the Divine Laws, to be the measure of the power that is in thee, and which this Verse would have thee employ for thy friends by commanding thee not to break easily with them and not to hate them for a slight offence.

For this Verse makes but little account of whatever does not affect the soul. It forbids us to make an enemy of a friend for the sake of self-interest and mercenary ends, and it commands us to endeavour, by an entire indifference to all exterior things, to regain our friend and to behave so that the whole world may bear witness to us, that as far as possible we have preserved our friend; that we have reclaimed and set in the right way those who had suffered themselves to be seduced by vice; that we have not given them any cause to break with us nor retaliated like for like when they first disclaimed our friendship.

The sacred Law of Friendship requires this of us — a law that may be said to be the most excellent of all virtues and to outshine all the rest in perfection — for the end of all virtues is friendship, and their principle is piety. The rules of piety are to us the seeds of all true goods, and the habit of friendship is the most perfect fruit of virtues.

As, therefore, we ought always to deal justly not only with those who are just to us, but likewise with those who endeavour to injure us (for fear lest by rendering them evil for evil we should fall into the same vice), so we ought likewise to have friendship — that is to say, humanity and good-will — for all that are of the same nature with us.

Now the due measure and regulation of friendship consists in loving, in the first place, the good and virtuous, as well for the sake of Nature as for the love of their inclinations (it being they alone who preserve in themselves the perfection of human nature), and in loving, in the next place, the wicked whose inclinations and principles have nothing in them that can induce us to court their friendship — in loving them, I say, for the sake of Nature only, which is common to them and us.

And therefore it is a true saying that *the wise man hates no one, but loves only the virtuous.* Inasmuch as he loves man, he hates not even the wicked, and inasmuch as he courts the virtuous to communicate and impart themselves to him, he selects above all the most perfect for the object of his affections. . . .

VERSES 9 AND 10

Know that all these things are as I have told thee.

Accustom thyself to surmount and vanquish these passions: First, gluttony, sloth, lust and anger.

These are the passions we ought to restrain and govern, that they may not discompose and obstruct our reason. To prevent this, let us gain the mastery of all our wild and brutish desires by good instructions, since their different parts reciprocally supply one another with arms to make us commit sins successively and, as it were, by degrees. For example, excess in eating provokes much sleep, and both together occasion a vigorous health which excites to lust and, provoking the concupiscible part of the soul, eggs it on to intemperance. At length the irascible part coming to join the concupiscible dreads no danger, is startled at no opposition, but dares undaunted the worst that can happen — to glut its depraved appetite, sometimes for luxurious eating, sometimes for carnal pleasures and sometimes for other delights. *Accustom thyself to surmount and vanquish these passions,* beginning with gluttony, that the irrational parts of the soul may accustom themselves to be obedient to reason, that thou may'st be an inviolable observer of piety to the Gods, of respect to thy parents and of all the other precepts I have already given thee. For the observance of those first precepts depends on the keeping of these, and thou wilt infallibly transgress the first if the passions be not kept in subjection and obedience to reason. On the one hand, either anger will provoke us against our parents or concupiscence will excite us to set at naught their good advices; on the other hand, either anger will precipitate us into blasphemy or the desire of riches will plunge us into perjury. In a word, all ill things are caused by these passions when reason is too weak to keep them within their bounds.

Thus you see which are the sources of all impieties, of all broils that set families at variance, of the treacheries of friends, and of all the crimes that are committed in breach of the Laws, so that some of the wicked are forced to cry out with Medea in the Tragedy,

> I know these crimes will blacken my lost soul,
> But rage my weaker reason does control.

Others: —

> I know the crimes I'm going to commit,
> But vanquish'd reason does to lust submit.

Or thus: —

> On me thy good advice is thrown away;
> My captive soul
> Is bound with shameful chains, nor can, nor will obey.

Whoever is capable of Reason, being in a good disposition and fitly prepared to discern what is honest and honourable, is always watchful, always ready to obey the precepts of Reason when the unruly inclinations of his passions, like so many lumps of lead, threaten to drag him down into the abyss of vice.

We ought, therefore, to know our duties and accustom, as much as we can, our brutal and sensual faculties to be obedient to the Reason that is in us. For the passions being thus kept in subjection, Reason will be in a condition to observe inviolably the first precepts, concerning which we are told in this place: *Know that all these things are as I have told thee.*

Concerning the following precepts we are told: *Accustom thyself to surmount and to vanquish,* etc., to make us understand that the Intelligent part is governed by instruction and knowledge, and that the sensual or brutal part is guided by habitude or by 'formations', if I may use that term, which are in some measure corporeal. Thus men tame and train up beasts by means of habitude only.

The appetite, therefore, being habituated to content itself with a sufficient and reasonable quantity, renders the other passions of the body more moderate and anger less violent and boiling, so that we, not being rudely agitated and hurried along by the passions, may meditate in tranquillity on what we are obliged to do and thence learn to know ourselves and what we truly are, and to respect ourselves when we have attained to that knowledge. From self-knowledge we learn to avoid shameful actions, that is to say, all the evils that are called shameful, because they are indecent and unworthy to be committed by a Rational Substance. And of this we are now going to speak.

VERSES 11 AND 12

Never commit any shameful actions, neither with others nor in private with thyself.

Above all things, respect thyself.

It frequently happens either that we commit shameful actions in private by ourselves for the shame of having a witness, or, on the contrary, that with others we commit those crimes that we should never have committed alone or in private, drawn in by company and the number of accomplices lessening the shame of the action. This is the reason why the Poet in this place counsels against these two ways that may lead us to shame and wickedness, for if all that is shameful ought to be avoided, no circumstance whatever can make it worthy of our search. Therefore he joined these two together — *neither with others nor in private by thyself* — to the end that neither solitude might induce thee to commit any indecency, nor company seem to thee to justify the crime.

After this he adds the cause that alone prevents from committing wickedness — *Above all things, respect thyself* — for if thou gain a habit of respecting thyself, thou wilt always have at hand a faithful guardian whom thou wilt respect, who will never depart far from thee but always keep thee in sight. For it has often happened that many, after their friends and domestics had left them, have taken the liberty to do such things which they would have been ashamed to have done in their presence.

Was there, then, no witness to it? I speak not here of God, for God is far from the thoughts of the wicked. But had they not their souls, that is to say, themselves, for witnesses? Had they not the testimony of their own consciences? Doubtless they had, but being subjected to their passions and enslaved by them, they knew not that such witnesses were present. And all who are in this condition condemn their own reason and treat it worse than the vilest slave.

Establish thyself, then, for thine own guard and thine own inspector, and keeping the eyes of thine understanding always fixed on this faithful guardian, begin to have an abhorrence of vice.

The respect thou shalt have for thyself will of necessity create in thee an abhorrence of all vice and incline thee to shun and avoid whatever is shameful and unworthy to be committed by a Reasonable Being. He who thinks ill actions unworthy of him unwittingly familiarizes himself with virtue. The Poet therefore goes on:

VERSES 13, 14, 15 AND 16

In the next place, observe Justice in thy actions and in thy words; and accustom not thyself to behave thyself in anything without rule and without reason.

Always make this reflection, that it is ordained by Destiny for all men to die; and that the goods of fortune are uncertain. As they may be acquired, they may likewise be lost.

Whoever respects and reveres himself becomes his own guard to prevent himself from falling into any manner of vice.

Now there are several sorts of vices: the vice of the rational part is folly; of the irascible, cowardice; the vices of the concupiscible are intemperance and avarice; and the vice that extends itself through all the faculties is injustice.

To avoid, therefore, all these vices we have need of four virtues: of prudence for the rational faculty; of courage for the irascible; of temperance for the concupiscible; and for all these faculties together we have need of justice, which is the most perfect of all the virtues, and being the chief of all, includes the rest as its proper parts. Therefore these Verses name justice first, next prudence, and after prudence they mention the most excellent effects that spring from that virtue and contribute to the perfection and to the entireness or totality of justice.

Every man who reasons right and who makes use of his prudence is assisted by courage in all good and praiseworthy actions, by temperance in the things that please the senses, and in each by justice. Thus prudence is the principle of all virtues and justice their end, and in the middle are courage and temperance. For the faculty that weighs and considers all things by right reasoning and that seeks out what is right in every action, to the end that all things may be done with reason and in due order, is the habit of prudence, that is to say, the most excellent disposition of our Rational Being by which all the other faculties are kept in good order, so that anger is brave and cupidity temperate. And justice, by correcting and amending all our vices and by animating all our virtues, adorns our mortal man with the excessive abundance of the virtue of the immortal man.

It is originally from the Divine Spirit that the virtues radiate and diffuse themselves in the Reasonable Soul; it is they that constitute its form, its perfection and all its felicity. And from the soul these virtues shine with a reflected ray on this senseless being — I mean the mortal body — by a secret and hidden communication, to the end that all that is joined to the Rational Essence may be filled with beauty, with decency and with order.

Now prudence, the first and as it were the guide of all Divine Goods, being established and firmly rooted in the Reasonable Soul, advises us well and makes us take the right course on all occasions, enables us to bear death with constancy and the loss of the goods of fortune with patience and mildness. For prudence alone can wisely and fearlessly support the vicissitudes of this mortal nature and of fortune that depends upon it. And indeed it is prudence that knows, by the help of reason, the nature of a thing and that it is of absolute necessity that what is composed of earth and of water should resolve into the same elements that compose it. Prudence never quarrels with Destiny, nor, because the mortal body dies, concludes therefrom that there is no such thing as Providence. She knows that it is ordained by Fate for all men to die; that there is a time prefixed for the duration of this mortal body; and that when our last moment is arrived, we ought not to repine, but submit voluntarily to it as to the Law of God. For that Law is properly what is meant by the word *Destiny*. It signifies that God Himself has by His decrees destined and prescribed necessary limits to this mortal life beyond which no man can pass, and it is the nature of prudence to follow the decrees of God, not by desiring not to die, but by endeavouring to die well.

Moreover, prudence is not ignorant of the nature of the goods of fortune. She knows that they come today and are gone tomorrow, according to certain causes that are predestined and prescribed and resistance to which is vain and dishonourable, for we are not the masters to keep what is not in our power.

Now most certainly neither our bodies nor the goods of fortune, in a word, nothing that is divided from our Rational Being, is in our power. And as it does not depend on us to get them when we please, so it does not depend on us to keep them as long as we please. But to receive them when they come, and to part with them when they go, and always to receive and part with them with constancy and virtue, is what depends on us and is the nature of our Rational Being. Without this it will accustom itself to a comportment without rule and without reason in all the accidents of life. We should form the habit of conforming to the orders of God, who has preordained and determined all that relates to us.

The main stress and extent, therefore, of what depends on us and is in our power, appears chiefly in this: that we can judge aright of the things that do not depend on us, and that we need not suffer ourselves to be deprived of the power of our free-will by an inordinate affection for frail and perishable things.

What is it that wise and sound judgement dictates to us? It bids us make good use of our body and of our riches whilst we have them, and make them serve as instruments and means to acquire virtue. And when we are on the point of losing them, it bids us rest satisfied of the necessity of it, and add to all our other virtues that of tranquillity and indifference.

The only way to preserve piety towards the Gods and the due proportion of justice is to habituate our reason to make good use of all the various accidents that arrive and to oppose the precepts of prudence to all the contingencies that seem fortuitous and to happen without order. For we can never preserve virtue unless our judgement be sound and our opinions wholesome. Never will the man who has accustomed himself to live without rule and without reason imitate the Beings that are better than us; but he will regard them as tyrants that force and constrain him. Never will he have the least regard for those amongst whom he lives, and never will he make good use of his body or estate.

Consider those who fear to die and who are wedded to their riches. See into what injustices, into what blasphemies, they necessarily plunge themselves by setting up the standard of impiety against God, by denying His Providence when they see themselves fallen into the things they foolishly thought to avoid, and by doing all sorts of injustices to their neighbour without making any scruple of ravishing from his estate that they may convert to their own use all they can unjustly scrape together! Thus it is manifest how these reprobates are misled by false opinions from which proceed the greatest evils — injustice to their equals and impiety to those above them. Only the man who, obeying this precept, is undaunted at the approach of death, whose judgement is refined and purged by Reason, and who does not believe the loss of temporal goods to be insupportable, is exempt from these evils. From these spring all the inducements and motives that incline him to virtue, for thence he learns that he ought to abstain from what is another's, to do no wrong to any man, and never to seek his own profit to the detriment and loss of his neighbour.

Now this can never be observed by the man who believes his soul to be mortal and who, having got a habit of living without rule and without reason, does not distinguish between what is mortal in us and has need of riches, and what is susceptible of virtue and is assisted and strengthened by virtue. For this due discernment only can incline us to the practice of virtue and stir us up to the pursuit of what is good and honourable. In this pursuit we are furthered and helped forward by a divine motive which springs from these two precepts: "Know thyself" and "Respect thyself." For our own worth and dignity ought to be the standard by which we should measure all our duties, both in our words and in our actions.

The observance of our duties is only the exact and inviolable observance of Justice. Thus you see that Justice is here placed at the head of all the other virtues so that it might be the measure and rule of our duties.

Observe Justice, says he, *in thy actions and in thy words.*

Let no blasphemy, then, proceed from thy mouth — neither for the loss of temporal goods nor in the sharpest pangs of a disease — that thou may'st not offend Justice in thy words. And never defraud thy neighbour of his goods, nor contrive mischief or loss to any man, that thou may'st not offend Justice in thy actions. For so long as Justice keeps, as it were, guard in our soul to protect and defend it, we shall perform all our duties towards the Gods, towards men and towards ourselves.

Now prudence is the best rule and the best measure of Justice. Therefore after the precept, *Observe Justice,* he adds, *and accustom not thyself to behave thyself in anything without rule and without reason,* seeing that Justice cannot subsist without prudence. And indeed there is nothing truly just but what perfect prudence has ordered. It is prudence that behaves in everything with Reason and examines and weighs with care what this mortal body is, what it has need of and what is necessary for its accommodation. It is prudence that takes everything to be vile and contemptible in comparison with Justice, and that makes all its good to consist in the best disposition of the soul which gives to all other things all the ornament and value they are capable to receive.

Thus you see the design of these Verses is to produce in the soul of those that read them these four practical virtues, with an exact and watchful observance of them both in word and deed. For one of these virtues inspires prudence, another courage, a third temperance, and that which precedes the three exhorts us to practise Justice, which diffuses itself through all the other virtues.

And this Verse: *The goods of fortune are uncertain. As they may be acquired, they may likewise be lost,* is here added to make us understand that the habit of temperance is generally accompanied with liberality, a virtue that regulates the receipt and expenditure of the goods of fortune. For to receive and expend them, when reason requires it, alone destroys meanness and prodigality.

All these virtues proceed originally from this principle of respecting oneself. And this precept "Respect thyself" is included in this, "Know thyself", which ought to precede all good actions and all other knowledge.

In effect, how can we otherwise come to know that we ought to moderate our passions and to understand the nature of things? For it is very much questioned, first, whether it is possible for man to do so, and in the next place, whether it is useful. On the contrary, good men seem to be much more unhappy in this life than the wicked because they take not unjustly from any man what they ought not to take, and because they pay every man what is his due. Moreover, with regard to the body, the good man is more exposed to ill-usage because he seeks not after rule and dominion and does not slavishly court those that govern.

If there is not in us an Essence whose sole delight and advantage is derived from virtue, it is vain for us to despise riches and honours. Thus you see why they who, being of opinion that the soul is mortal, teach that we ought not to abandon virtue, are rather idle talkers than true philosophers. For if there subsists not something of us after death, and if that something too were not of a nature capable of being adorned with virtue and with truth, such as we believe the Reasonable Soul to be, our desires would never be fixed on good or honest actions. The bare suspicion of the soul's being mortal stifles and deadens in us all our zeal for virtue and excites us to the enjoyment of corporeal pleasures, whatever they be. . . .

VERSES 17, 18, 19 AND 20

Concerning all the calamities that men suffer by Divine Fortune, support with patience thy lot, be what it will, and never repine at it, but endeavour what thou canst to remedy it, and consider that Fate does not send the greatest portion of these misfortunes to good men.

Before we enter any further on the explanation of these Verses, it is requisite to take notice that what the Poet here calls calamities are all the pains and afflictions that render this life most difficult, laborious and uneasy, as diseases, poverty, the loss of the friends and acquaintances that are most dear to us, and to be condemned in our country and the like. All such things are troublesome and hard to bear; they are not, however, real ills, and they hurt not the soul unless it suffer itself to be led by them into vice, which would equally ensue from the enjoyment of good things if we refused to make good use of them, whether they be health, riches or dignities. For we may be depraved by these, as we may be sanctified by their contrary.

Now the real ills are the sins we commit voluntarily and of our own free choice, and which are incompatible with virtue, such as injustice, intemperance, and all the other things that can in no way be united, matched or reconciled with what is good, well, fine or decent. It is not possible on account of any of these vices to cry out "This is well", or the like. For example, no man will ever say "It is well or good to be so unjust; it is well done to be so intemperate." On the contrary, we daily say of exterior evils, "It is well done to bear sickness in such a manner; it is well done to undergo poverty like such a one", when any man supports these accidents with constancy and according to the dictates of right Reason. But these exclamations can never be congruously applied to the vices of the soul because they are deviations from right Reason and contradictions to it, which, though it be naturally imprinted in the souls of all men, is nevertheless imperceptible to a man blinded by his passions.

Now a certain proof that right Reason is naturally in man is that even an unjust man, when his interest is not concerned, judges with Justice, and an intemperate man with temperance, and in a word, that the wicked have good motions and inclinations in all the affairs that concern them not and wherein their passions do not overrule them.

This is the reason that all vicious men may amend their lives and become virtuous if they condemn and forsake their former vices. It is not of necessity that there should be a pretended extravagant reason as the principle of vice, as right Reason is the principle of virtue, for right Reason is sufficient for all, as the laws of a country suffice to determine what is done according to them or against them, and to approve the one and condemn the other. There is no need to set up a principle of ill, whether we make it proceed from within or from without us. We need only the principle of Good, which by its Essence is separated from Rational Substances, and this principle is God. But it is likewise within those Substances and governs them by its power, and this is right Reason.

Let us now observe the difference the Poet makes between evils.

Speaking of voluntary evils, he does not say that they are distributed by *Divine Fortune*, but he says it of the ills that are exterior and conditional, that do not depend on us in this life, but are the effects of the sins we formerly committed. These ills are indeed painful and calamitous, as we have said already, but they may receive a lustre and ornament from the hands of virtue. For a regular and temperate life gives lustre to poverty; prudence ennobles a mean extraction; the loss of children is alleviated by a due submission that makes the father say, "My son is dead, and thus I have returned back what was lent me", or "I knew he was conceived in mortality."

In like manner, all other ills, when illustrated by the presence of virtue, become resplendent and even worthy of envy.

Let us now enquire what is meant in these Verses by *Divine Fortune* by which men fall into outward evils, for if God gives beforehand and of Himself — to one riches, to another poverty — this should be called *Divine Will*, not *Fortune*. And if nothing governs or presides over these dispensations, but if it be true that these ills are fortuitous and arrive by chance, and thus one man comes to be happy, as we call it, and another unhappy, this ought to be called *Fortune* only, and not *Divine Fortune*.

But if God, who takes care of us, distributes to each man according to his merits, and if He be not the cause of our wickedness but only the Master to render to each according to his works by following the sacred Laws of Justice, then it is with good reason that the Poet gives the name of *Divine Fortune* to the manifestation of His judgements. The Poet, inspired by the God that dispenses these judgements, has put the epithet *Divine* first, and since those whom God judges have deprived themselves voluntarily and through their own free choice (and thereby have justly deserved His punishments), he adds to the epithet the substantive *Fortune* because God never intends to punish or reward men beforehand, but to treat them according to what they are after they have become that which they themselves are the causes of.

This mixture, therefore, and this alloy of our will and of His judgement, is what produces that which he calls *Fortune,* so that the whole together, *Divine Fortune,* is only the manifestation of the judgement of God against sinners. And thus the ingenious connection of these two words joins and links together the care of God who presides and the liberty and free motion of the soul that elects, and lets us see that these ills arrive neither absolutely by the decrees of Destiny nor by the orders of Providence, not fortuitously nor by chance. It is not our will alone that disposes of and determines all the actions of our life; but all the sins we commit are imputed to our will, and all the punishments that follow these sins, according to the Laws of Justice, are ascribed to Destiny. The good things that God bestows beforehand and without our previous merit are attributed to Providence, for nothing that exists is referred to chance. . . .

If there were no Providence, there would be no Order in the world, and this Order may be called Destiny. If there were neither Providence nor Order, there would be neither Judgement nor Justice, nor even so much as rewards and honours for good men. But there being a fixed and certain Providence and Order, all men who are born into the world must necessarily have a like share in all the same goods, unless they contribute to what causes the inequality. It is manifestly evident that all good things are equally distributed, and it is equally evident that the inequality and unlikeness of men's wills, which is submitted to the judgement of Providence, prevents them from having the same share, since their lot must of necessity be proportioned to their merit.

Let us not concern ourselves that the same inequality reigns as well in the brute beasts, in plants and in all inanimate things as in man. Because we see that chance rules over all these things that are so inferior to man, we should not conclude that Providence does not watch over us; nor because whatever relates to us is absolutely determined and brought to pass, should we conclude that the Justice and Judgement which God extends over all these inferior Beings is likewise in them a token and effect of their virtue or their vice. The things that are merely inanimate serve as common matter to plants and to animals, and moreover the plants serve as nourishment to men and to beasts, some of which last are destined to nourish their fellow beasts and men. It is, therefore, evident that these things are not thus ordered and done with regard to the merits of any of those animals, but proceed from their endeavour and from a desire they have to satisfy their hunger, to heal their diseases, or in a word, to relieve their several wants and necessities in the best manner they can (insofar that the unhappy condition of beasts is occasioned by our necessities which they are destined to relieve). On the contrary, the cause of what we call their happiness is the affection which we sometimes allow ourselves to feel for them. . . .

VERSES 21, 22 AND 23

There are amongst men several sorts of reasonings, good and bad. Admire them not too easily and reject them not either, but if any falsehoods be advanced, give way with mildness and arm thyself with patience.

The will of man not persisting always in virtue nor being always bent upon vice produces the two sorts of reasoning and discourse that bear the marks of those two contrary dispositions in which it successively is. Hence it comes to pass that of those reasonings, some are true, others false, some good, others bad.

This difference requires on our part a just discretion which is the result of knowledge, that we may choose the good and reject the bad, and likewise that we may not fall into a hatred of all ratiocinations because they contain some bad arguments that we condemn, nor admit them all as good without distinction because there are some good ones that we receive. By the hatred of all reasonings in general we deprive ourselves of those that are good, and by admitting all without distinction we expose ourselves to be misled by the bad.

Let us, therefore, learn to love reasonings, but with a just discernment, to the end that the love we have for them may make us willing to hear all, and our discretion allow us to reject those that are bad.

In doing this we shall observe the precept of Pythagoras. We shall not admire the reasonings that are false, nor admit them without examination under pretext that they are reasonings, nor deprive ourselves of those that are good under pretext that all reasonings are bad. . . .

VERSES 24, 25 AND 26

Observe well, on every occasion, what I am going to tell thee: Let no man either by his words, or by his actions, ever seduce thee, nor entice thee to say or to do what is not profitable for thee.

This is a precept of general extent and much the same as that already given in the eleventh and twelfth Verses: *Never commit any shameful actions, neither with others nor in private with thyself. Above all things, respect thyself.* The man who neither alone nor in company will dare to commit the least shameful action, but out of respect for the Reason he has within him and to whose government and conduct he has resigned himself, banishes the very thoughts of such actions, he alone, I say, is in a condition to obey this precept: *Let no man either by his words, or by his actions, ever seduce thee.* He alone is incapable of being cheated and misled who, having a due sense of his own nobility and dignity, does not allow himself to be cajoled by flatteries nor intimidated by threats, whatever means are used for that purpose either by his friends or by his enemies. The words *no man* include all men, whether they be a father, a tyrant, a friend, an enemy. And the different methods of deception proceed either from words or from actions: from the words of those that flatter or threaten, and from the actions of those that offer bribes or that set before us pains and punishments.

Against all these things, therefore, let us keep our soul well strengthened and fortified by sound Reason, to the end that it may neither be enslaved by any accidents that can happen from without, whether delightful or painful. For sound Reason, having established temperance and fortitude in the soul as two guards that are always watchful and cannot be corrupted, will always preserve us from being ever seduced either by the charms of delights or by the dread of things that are terrible. This produces the exact Justice which the Poet has already commanded us to observe in our words and in our actions.

By this means no man, whoever he be, will ever prevail upon us to commit the least action, or to let drop the least expression, that is not consonant with right Reason. For if we respect ourselves above all things, no man will appear to us to be more worthy to be respected or feared than ourselves.

Therefore thou oughtest thoroughly to understand this saying, *what is not profitable for thee*, and refer this pronoun *thee* to what thou truly art. For if thou understandest this precept aright, *Let no man either by his words, or by his actions, ever seduce thee, nor entice thee to say or to do what is not profitable for thee*, and if thou, properly speaking, art the Reasonable Soul, thou wilt never suffer, if thou art wise, anything that can be harmful to thee — thee, I say, who art a Rational Being — for thou art properly the soul.

The body is not thee, it is thine; and all exterior things are neither thee nor thine, but they belong to something that is thine, that is to say, to thy body.

If thou distinguish and divide in this manner all these things, thou wilt never confuse them. Thou wilt discover what the Essence of man truly is, and by not taking it either as the body or the things exterior to the body, thou wilt not trouble thyself for this body nor for what belongs to it, as thou wilt do for thyself, to the end that such a mistaken care and concern may not seduce thee into a love of the body and of riches.

Whilst we are absolutely ignorant of what we are ourselves, we shall be ignorant likewise of the care we ought to take of things, and shall rather take care of anything rather than of ourselves, which nevertheless ought to be our chief concern.

And indeed, if the soul is that which makes use of the body, if the body serves as an instrument to the soul, and if all other things were invented in favour of this instrument and for the support of its nature which daily decays and perishes, it is clear that our first care ought to be for that which is first, and our next care for that which holds the second rank. The wise man, therefore, will never neglect his health — not that he gives the first rank to the body or takes it for his principal part — but that he may preserve it in a condition to supply all the wants of the soul and to obey all its orders without hindrance. And lastly, his third care will be for what is third in order, and he will govern with prudence and economy all exterior things that serve for the preservation of the instrument, which is his body.

Thus his chief, or to say better, his only care, shall be for his soul, seeing that the care we have of other things, next to the soul, is only for the sake of the soul and tends alone to its preservation and profit.

Now whatever is foreign to virtue is what this Verse here expresses in these words: *what is not profitable for thee.* If virtue is profitable for thee, whatever is not virtue will be useless — nay, pernicious to thee. We are therefore advised to throw up, as it were, a rampart around us for the preservation and defence of virtue by him who tells us that we ought never to obey those that use all their powers to make us swerve from virtue. . . .

VERSES 27, 28 AND 29

Consult and deliberate before thou act, that thou may'st not commit foolish actions, for it is the part of a miserable man to speak and to act without reflection.

But do that which will not afflict thee afterwards, nor oblige thee to repentance.

Wise and prudent deliberation is the mother of virtues and perfects and preserves them, as it is their mother, their nurse and their guard. When we consult quietly within ourselves what course of life we ought to follow, we choose virtue for the sake of its beauty.

After this choice the soul, strengthened by this consultation, encounters and supports all toils and conflicts on account of virtue. Because it has already become accustomed to honest and excellent things, it preserves its judgement sound and entire amidst the worst calamities. Nor can anything that comes from without ever cause it to change its opinion that there can be any other happy course of life but that which of its own free will it chose and embraced, after having judged it to be the best and most excellent.

Hence it comes that there are three sensible effects of wise deliberation: First, the choice of the best life; secondly, the practice of the life we have chosen; and thirdly, a constant and exact observance of what we had well and wisely resolved upon.

The first of these effects is the reason that precedes the execution of what we desire to do and defines the principles of the action. The second is the reason that accompanies the execution and adjusts beforehand each particular action to the principles that precede it. And the third is the reason that follows the execution, examines each action we have done, and judges whether it was well done and as it ought to be.

In all things whatever, the shining beauty of wise and prudent deliberation is eminently conspicuous. Sometimes it produces virtues, sometimes it nourishes and perfects them, and lastly, it is watchful to preserve them, so that it is itself the beginning, the middle and the end of all good things.

In wise deliberations we find a deliverance from all our ills, and by it, and it alone, are we enabled to bring the virtues to perfection. Since ours is an Intelligent nature, and consequently capable of consulting and deliberating wisely, if we choose well, the good life we embrace preserves its Essence untainted; whereas if we make a choice without reason, that choice corrupts it as much as in it lies.

Now the corruption of what is immortal is vice, the mother of which is *temerity*, which this Verse commands us to avoid, *that we may'st not commit foolish actions.* The foolish actions are the wicked and ill actions, for *to speak and to act without reflection is the part of a miserable man*, that is to say, is the nature of a wicked man. But if thou dost deliberate before thou act, thou wilt never commit any of these foolish actions which can only serve to afflict thee afterwards. Repentance evidently demonstrates the badness of the choice of which experience has shown thee the disadvantage. On the contrary, the effects of good consultation are evidence for the goodness and of the safety of the choice by proving, even by the very actions themselves, the advantages that result from them. The advantages I speak of are not in relation to the body nor to any exterior things, but to ourselves — the advantages *that will not afflict our souls.* For what advantage is it to a man to heap up vast riches by perjuries, murders and by all sorts of other ill actions? What will he gain by exterior riches if he leave his soul in poverty, without the only goods that can be useful to it, and to be, besides, reduced to the wretched state of insensibility which increases his misery; or, if his conscience awaken in him a sense of his crimes, to suffer unspeakable tortures in his soul which result from that remorse, to be night and day in continual dread of the punishments of Nemesis, and to find no other remedy to his ills than to have recourse to thoughts of annihilation? . . .

We ought, therefore, above all things to endeavour not to sin at all. But when we have sinned, we ought to embrace the punishment as the sole remedy for our sins and as what will correct our rashness and our folly. For the innocence we lost by sin we recover by repentance and by the good use we make of the punishments with which God chastises us.

Repentance is the beginning of Philosophy, the avoiding of all foolish words and actions, and the first step of a life that will no more be subject to repentance. He who prudently deliberates before he acts never falls into involuntary and unforeseen troubles and misfortunes nor ever commits unwittingly any actions whose consequences he can foresee. But he prepares himself to accept whatever can happen contrary to his expectation.

Therefore, neither the hope of what we call goods makes him renounce his real good, nor does the fear of evils incline him to commit real ills. But having his mind continually bent on the rules that God has prescribed, he squares his whole life according to them.

That thou may'st know most assuredly that it is the part of a miserable person indeed to speak and to act without Reason, behold Medea deploring her miseries in our theatres. The fury of a senseless amour spurred her on to betray her parents and to run away with a foreigner. At length, finding herself condemned and forsaken by him, she thought her misfortunes insupportable, and in thought breaks out into imprecation, "Let Heaven's dire thunder on my head be hurled!" after which she falls to committing the most heinous of crimes.

In the first place it is unreasonable and foolish for her to pray that what is done might be undone; and then, like a senseless distracted person indeed, she thinks to heal her ills by other ills, hoping to efface the beginning of her miseries by a yet more miserable end. For she madly endeavours, by the murder of her children, to atone for her marriage, to which she had consented rashly and without reflection.

If you have a mind to see how Homer's Agamemnon behaves, you will find that prince, when he is punished for not having bridled his rage, crying out with tears in his eyes, "I'm lost! undone! and all my strength forsakes me!" And in the ill state of his affairs he quenches with a flood of tears that fire of his eyes which rage had kindled in his prosperity.

This is the life of every foolish and inconsiderate man. He is driven and tossed to and fro by contrary passions, unendurable in prosperity, dejected in adversity, imperious and haughty when he hopes, cowardly and crouching when he fears. Not having the constancy that prudent deliberation inspires, he veers about with every blast of Fortune.

Let us then take sound Reason for our guide in all our actions, imitating Socrates, who says: "You know that I have now accustomed myself not to obey any of mine except that Reason which after due examination appears to be most just and upright." By this expression "any of mine", he means all his senses. And indeed, all the things that are given us to be subservient to Reason — as anger, love, sense, and even the body itself, which is to serve as an instrument to all these faculties — are *ours*, but not *us,* and we ought to obey none of them except, as Socrates says, sound Reason alone — that is to say, the Rational part of our nature. It is that alone that can see and know what ought to be done and said.

Now to obey sound Reason and to obey God are the same thing, since our Intelligent part is enlightened by the irradiation that is natural and proper to it and wills nothing but what the Law of God requires. A soul well disposed according to God is always of the same mind with God, and whatever it does, it keeps the divinity and splendid brightness that surround it always in sight;

whereas the soul disposed in a contrary manner, intent on what is out of God and full of darkness, is carried here and there and wanders without keeping any certain road, being destitute of understanding and fallen from God.

These are the great and infinite advantages that arise from prudent and wise deliberation, and the great mischiefs that necessarily follow temerity and want of reflection.

But besides all these great advantages of which we have spoken, *to consult before we act* produces one more vast importance, which is that it checks all the motions of opinions, brings us to the true knowledge of things, and makes us lead a life that cannot fail to be most pleasant, since it must be most just and good, as will appear by what follows.

VERSES 30 AND 31

Never do anything which thou dost not understand; but learn all thou oughtest to know, and by that means thou wilt lead a very pleasant life.

Refraining from the things we do not understand will hinder us only from committing errors; but to learn that which leads to a good life not only prevents us from making faults, but directs our actions and gives us success in our undertakings. The knowledge of our own ignorance curbs the temerity that opinion excites, and the acquisition of knowledge secures the success of all our enterprises.

Now these are two excellent things: to know that we do not know, and to learn what we are ignorant of. They are followed by the best and most happy life. But this happy life is only for him who is free from opinion and replenished with knowledge, who is not puffed up with vanity on account of anything that he knows, and who desires to learn whatever deserves to be learnt. Now nothing deserves to be learnt except that which brings us to the Divine Likeness; inclines us to deliberate before we act; puts us in a condition not to be deceived and misled by any man, either by his words or by his actions; enables us to discern the difference in the reasons and arguments which we hear; makes us bear with patience the *Divine Fortune* and that supplies us with means to mend it; teaches us not to dread death and poverty, and to practise Justice; makes us temperate in all things that are called pleasures; instructs us in the Laws of Friendship and the respect due to those that gave us life; and lastly, shows us the honour and the worship we ought to render to the superior Beings.

These are the things that this Verse tells us we ought to learn and that they are attended by a most pleasant life. For he who distinguishes himself by his virtue enjoys pleasures that are never followed by repentance and that imitate the solidity and permanence of the virtues that procure them. For all pleasure is naturally the consequence of some action, whatever it be. Pleasure subsists not by itself, but arrives when we do such and such an action. Hence it is that pleasure always follows the nature of the action that produces it. The worst actions produce the worst pleasures, and the best actions produce likewise the best pleasures, so that the virtuous surpass the vicious not only in regard to the beauty of their actions, but for the degree to which alone the vicious seem to have plunged themselves into vice.

And indeed, as much as one disposition is better than another disposition, so much, too, is one pleasure preferable to another pleasure. Therefore, seeing that a virtuous life in which visibly appears the likeness of God is truly divine, and seeing that a vicious life is brutal and without God, it is evident that the pleasure of the virtuous imitates and approaches the Divine Pleasure in following the dictates of the understanding and even God Himself. And the pleasure of the vicious (for I am content to call them both by the same name) imitates only the brutal gratifications of a sensual appetite, the delight of beasts.

Pleasures and sorrows change us and alter our condition. Pleasures are placed within our reach, and he that enjoys them when, where and as much as he ought, is happy. He that knows not these just bounds is unhappy. Therefore, the life void of opinion is exempt only from sin, and the life that is full of knowledge is always happy and perfect and by consequence the best and the most delightful.

Let us, then, never do what we know not how to do, and as to what we understand let us do it when we ought. Ignorance produces faults, and knowledge seeks *the opportunity*, for many things that are very good in themselves become bad when they are done out of due season. Let us, then, obey this precept in order: as it commands us to check and keep back our actions, it endeavours to render us free from faults; and as it commands us to learn, not everything but what deserves to be learnt, it excites us to honest and virtuous actions. A good life consists not in being exempt from faults, but in doing all that ought to be done. For the first it is sufficient to purge the opinion, but the last can be nothing but the effect of knowledge.

By living exempt from faults and by living a good life, behold the advantage that will accrue to thee: *Thou wilt lead a very pleasant life.*

What is this pleasant life? The life that enjoys all the pleasure that arises from virtue and in which the good and the delightful meet each other. If, therefore, we desire what is good, and at the same time what is pleasant, what will such a mixture be but what this Verse tells us, a most pleasant life? For he that chooses the pleasant with the shameful, though for a little while he enjoys the charm of the pleasure, yet the persistence of shame in the enjoyment will soon throw him into the bitterness of repentance. Whereas he that chooses the good with the painful, though at first the task sit heavy on him, the good will soon alleviate and lessen his toils, and in the end he will enjoy with virtue all the fruits of pure and unmixed delight.

To conclude, if we do any shameful thing with pleasure, the pleasure passes but the shame remains. But if we do any good thing with a thousand toils, a thousand difficulties, the toils and difficulties will all vanish and be forgotten and the good alone will remain with us. Hence it necessarily follows that an ill life is very sorrowful and troublesome and a good life is most delightful and pleasant.

Let this suffice for the understanding of these Verses. But seeing the care of the body conduces to the perfection of the soul, let us see what the Poet next adds.

VERSES 32, 33 AND 34

In no wise neglect the health of thy body; but give it food and drink in due measure, and also the exercise of which it has need. By measure, I mean what will not incommode thee.

This mortal body having been given us as an instrument for the life we are to lead here below, we ought neither to pamper it by too indulgent a treatment nor pinch and bring it low by too austere and sparing a diet. Both produce the same impediments and hinder the use we ought to make of it. Therefore we are here exhorted to take moderate care of it and not to neglect it, either when it is grown rebellious by too high feeding, or when it is mortified and brought down by sickness, to the end that being kept in the temper in which it naturally ought to be, it may perform all the functions that the soul which guides it shall require of it.

For the soul makes use of the body, and the body serves the soul. The workman, then, is obliged to take care of the instrument he employs. It is not enough only to desire to make use of it, but all the reasonable and necessary care must likewise be taken to keep it always in a condition to execute our orders.

Because it is naturally in a continual state of generation and of corruption, and seeing that repletion and evacuation entertain and nourish it — sometimes aliment making good and repairing what is wasted, and sometimes exercise evacuating and carrying off what is excessive — we ought to balance the nourishment that causes repletion with the exercise that causes evacuation. This *due measure* is the Reason that adapts the habit of the body to the intellectual operations of the soul, and which, by this means, takes such care of the health of the body as suits and becomes a philosopher.

This Reason, therefore, will make choice of such food and exercise as will not make the body too fat nor hinder it from following the intellectual aspirations of the mind. For it is not merely a body of which it takes care, but a body that is subservient to the thoughts of the soul. It therefore rejects the athletic regimen or source of life because that takes care only of the body without having any regard to the soul; it avoids all superfluous care of the body as being entirely contrary to the Intelligent Light of the soul. But the regimen of life which can most of all contribute to the learning of the Sacred Sciences and the performance of all good and honourable actions, is that which ought to be chosen by the man who wishes to embrace the life of Reason. To him these words are addressed: *By measure, I mean what will not incommode thee.*

Let not, then, the *measure* of the care thou takest of thy body incommode thee in the least, thou who art a Reasonable Soul, thou, who being an observer of all the precepts already given thee, art obliged to make use of such drink, food and exercise as will render the body obedient to the commands of virtue, and will not provoke the sensual and brutal part to be refractory and take head against the Reason that guides it. . . .

VERSES 35, 36, 37 AND 38

Accustom thyself to a way of living that is neat and decent, without luxury. Avoid all things that will occasion envy, and be not expensive out of season, like one who knows not what is decent and honourable.

Be neither covetous nor niggardly. A due measure is excellent in these things!

It is not only in drinking and eating that it is good to observe *a measure*, says the author of these Verses, but also in all other things. *Measure* is equally distant from too little and too much because in everything we may doubly exceed this *due measure*, either by being excessive or too niggardly, both of which are blameable, unworthy of a philosopher, and far from the medium we ought to observe in everything that relates to the body. Too much neatness begets luxury and effeminacy, and too much simplicity or plainness degenerates into meanness and slovenliness.

To avoid, therefore, falling into the first defect through too much neatness, or into the second through too much simplicity, let us keep the mean between them, declining the neighbouring vices of these two virtues and making both of them serve for a corrective remedy to each other.

Let us embrace a plain way of life so as not to be slovenly, and a neat way of life so as not to be over-refined and luxurious. Thus we shall observe the *due measure* in all that relates to the body: our apparel will be neat and clean but not costly and magnificent; our house and table will be neat but not splendid and luxurious; and let us behave ourselves in like manner as to our furniture and everything else. For seeing that the Reasonable Soul commands the body, it is just and decent that whatever relates to the body should be ordered by Reason which, being persuaded that everything ought to be answerable to its dignity, will suffer neither luxury nor sordidness. . . .

Let all things, therefore, necessary for life be tempered with so just a mean as equally to balance the two contrary extremes. *Accustom thyself,* says the Poet, *to a way of living that is neat and decent,* but perceiving afterwards that this neatness might throw us into luxury, he adds *without luxury.* He would only have said accustom thyself *to a way of living without luxury,* but he foresaw that such a simplicity might be apt to make us fall into sordidness. Therefore he joined these two together, *neat and without luxury,* to prevent our falling into the excess of either by the counterpoise of one against the other, to the end that from both of them might proceed a masculine sort of life such as becomes a Rational Creature.

By ordering our life in this manner we shall gain another great advantage by avoiding the envy that always attends extremes, since by running headlong into excess in all things, we provoke our neighbours sometimes to hate us for our luxury, sometimes to complain of our slovenliness, now to accuse us of prodigality, and then to reproach us for stinginess and meanness of soul. All these excesses make us alike incur the blame of those amongst whom we live.

And this is what the word *envy* in this place properly signifies, for in bidding us to *avoid all things that will occasion envy,* he means what justly will expose us to the blame of men.

Now Reason and the general voice of the world blame luxury and sordidness in the manner of living, and profuseness and niggardliness in economy. Decency, therefore, and moderation in all exterior things show the good disposition of our soul and let us see that a *due measure* is best in everything. For the man who loves his repose ought as much as possible to shun all occasions of envy and to be as much afraid of provoking it as of rousing a sleeping lion, to the end that without any disturbance he may advance in the study of virtue. . . .

VERSE 39

Do only the things that cannot hurt thee, and deliberate before thou doest them.

This is a precept he has already often given us, first in these words, *But do that which will not afflict thee afterwards,* and again, *By measure, I mean what will not incommode thee,* and in a third place, *Let no man either by his words, or by his actions, ever seduce thee, nor entice thee to say or to do what is not profitable for thee.* And here, by this short recapitulation, he sets again before our eyes all these precepts by advising us to abstain from everything that may hurt us and to do all that may be of use to us.

It is easy to distinguish between these two sorts of action if we deliberate before we act and reflect what ought to be done and what ought not to be done. The time for deliberation and consultation is whilst matters are yet entire and before we have set our hands to the work.

As to what he says in this place of *the things that cannot hurt thee,* we will explain it as we have already explained the precept he gave us before, when he said *that which will not afflict thee,* and say that by this *thee* he means that which is indeed man, the Rational Being, that is to say, the man who has embraced wisdom and who uses all his endeavours to render himself like God.

This inward man is wounded by whatever is contrary to right Reason, by whatever is contrary to the Divine Law, and by whatever hinders the resemblance with God and destroys His Image in us. All these things generally proceed from the conversation of those with whom we live, from the care we have of the body to which we are united, and from the use we make of riches which were invented only to be a help to the body and which for that reason are called by a name that implies that they ought to be made use of for the necessities of the body.

He, therefore, who is inflamed with the love of Divine Goods ought to take great care never to be prevailed on to do what is not useful for him, never to give his body what will be hurtful, never to receive or admit anything that can prevent him from the study of wisdom and for which he will soon have cause to repent.

We ought to prevent all these things by deliberating before we act, so that when we come to examine into all our past actions, we may remember them with pleasure and delight. And this is his design in the following Verses.

VERSES 40, 41, 42, 43 AND 44

Never suffer sleep to close thy eyelids after thy going to bed, till thou hast thrice reviewed all thy actions of the day: Wherein have I done amiss? What have I done? What have I omitted that I ought to have done?

If in this examination thou find that thou hast done amiss, reprimand thyself severely for it; and if thou hast done any good, rejoice.

Here thou should'st recollect in thy memory all the precepts already given thee, to the end that, regarding them as Divine Laws, thou may'st make a just examination in the inward tribunal of the soul of all thou hast done well or done amiss. For how can the enquiry into our past actions enable us to distinguish when to reprimand and when to praise ourselves, if the deliberation that precedes them had not placed before us certain laws and rules according to which we ought to regulate our life, and which ought to be in regard to us as a divine mark, according to which we are to examine all the recesses of our conscience.

Pythagoras requires us to make this examination daily, that by frequent and assiduous recollection our memory may be the more certain and the more infallible. And he will have us do it every evening before we go to sleep, to the end that each night, after all the actions of the day, we may give ourselves an exact account of them before the tribunal of conscience.

This severe examination of our dispositions may be sung as a hymn of praise to God at our going to bed: *Wherein have I done amiss? What have I done? What have I omitted that I ought to have done?* By this means the whole tenor of our life will be ordered according to the precepts that have been prescribed, and we shall conform our Reason that judges to the Divine Intelligence that made the Law.

What says the Legislator? That we ought to revere the superior Beings according to the order and rank of their Essence; have much veneration and respect for our parents and relations; love and embrace good men; keep in subjection our passions and worldly desires; respect ourselves everywhere and in all things; practise Justice; consider the shortness of life and the instability of riches; receive with submission the lot which the Divine Judgement sends us; take delight only in the thoughts that are worthy of God; keep our mind continually bent on what is most excellent; love and embrace only the reasons that truly deserve that name; put ourselves in a condition that we may preserve the precious storehouse of virtue; consult before we act, that repentance may not be the fruit of all we do; free ourselves from all opinion and obstinacy; seek after the life of knowledge and apply and adapt our body and all exterior things to the functions of virtue.

These are the Laws that the Divine Intelligence imposes on the soul; and no sooner has Reason received them than she becomes a very watchful guard for and over herself. *Wherein have I done amiss? What have I done?* says she every day, regularly calling to mind all her actions, good and bad. If at the end of this examination she find that she has passed the day without violating any of these Laws, she makes herself a garland of the fruits of divine joy; and if she catch herself in any crime, she then punishes herself by the severe correction of repentance, as by astringent remedies.

Thus you see, says the Poet, why you ought to banish sleep to give time to Reason to make this examination. The body will suffer little by being thus kept awake, because it has not made a habit of sleeping by reason of its prudent and temperate diet, a regimen by which even our most natural passions are subjected to the empire of Reason.

Never suffer sleep to close thy eyelids after thy going to bed, till thou hast thrice reviewed all thy actions of the day. What is this examination? *Wherein have I done amiss? What have I done? What have I omitted that I ought to have done?* For we sin in two manners: either by doing what we ought not to do, which is expressed in these questions, *Wherein have I done amiss? What have I done?* or in not doing what we ought to do, which is expressed word for word in this Verse: *What have I omitted that I ought to have done?* It is one thing to omit good, and another to commit evil. One is a crime of omission, the other of commission. For example: *We ought always to pray and never to blaspheme; we ought to nourish our father and our mother, and we ought never to use them ill.* He who does not keep the first two points of these two precepts does not do what he ought, and he who transgresses the last two does what he ought not, though it may be said that both these precepts are in some manner alike, seeing that they deal with the transgression of the same Law.

Thus the Poet exhorts us to make an examination into all the actions of the day, from the first to the last in order, without forgetting the intermediate actions, which is expressed in these words: *Continue to go on in this manner.* For it often happens that the transposition deceives the judgement and makes it favour some actions which, had the memory recollected in order, would have been inexcusable. Moreover, this recapitulation of the life we have led in the day refreshes in us the remembrance of all our past actions and awakens us to the thoughts of immortality. . . .

VERSES 45, 46, 47 AND 48

Practise thoroughly all these things; meditate on them well; thou oughtest to love them with all thy heart. It is they that will put thee in the way of Divine Virtue.

I swear it by Him who has transmitted into our souls the Sacred Tetraktys, the Source of Nature, whose course is eternal.

This is what I have already said in the Preface, that Practical Philosophy makes a man good by the acquisition of virtues, that Contemplative Philosophy makes him like God by the irradiation of understanding and of truth, and that in what relates to us, small things ought necessarily to precede the greater. For it is easier to conform human life to the rules of Reason than it is to incline it to what is most divine and most high. This cannot be done but by giving ourselves wholly up to contemplation.

Besides, it is impossible that we should enjoy truth undisturbed if our sensible faculties are not in entire subjection to the moral virtues, according to the law of the understanding. For the Rational Soul, holding the middle rank between the understanding and what is deprived of Reason, cannot inseparably adhere to this understanding which is above it until, being purified and freed from all affection for the things that are below it, it makes use of them with purity. It will be pure when it no longer allows itself to be seduced and led astray by what is void of Reason or by this mortal body, and takes no further care of it than of things that are foreign to it and what is permitted by the Law of God which forbids us in any manner to throw off our chains, but commands us to wait until God Himself comes to deliver us from our captivity.

Such a soul, therefore, has need of both sorts of virtues: Civil or Practical Virtues to regulate and moderate the rage of desire that inclines it towards the things here below; and Contemplative Virtues that incline and raise it up towards the things above and that unite it to the superior Beings.

Between these two virtues the Poet has put two Verses to be, as it were, the boundaries to divide them. The first: *Practise thoroughly all these things; meditate on them well; thou oughtest to love them with all thy heart,* is a very proper conclusion to Civil Virtue; and the last, *It is they that will put thee in the way of Divine Virtue,* is the beginning of Speculative Knowledge and, as it were, a noble entry that leads to it.

This beginning promises to him who has laid aside the sensual life, who has delivered himself as much as possible from the excess of passions, and who thereby has become man from man that he is, he shall commence to be God as much as it is possible for human nature to participate of the Divine Essence.

That this deifies us and is the end of Contemplative Truth is evident by these Verses which he puts at the end of this treatise, as a noble conclusion that leaves us no room to wish for more: *And when, after having divested thyself of thy mortal body, thou arrivest in the most pure Aether, thou shalt be a God, immortal, incorruptible, and death shall have no more dominion over thee.* It is of necessity that we shall obtain this happy re-establishment of our primitive state, that is to say, this glorious apotheosis, by the constant practice of virtues and by the knowledge of truth. This is what this sacred book evidently demonstrates to us, as we shall soon see.

Let us now return to the Verses at present before us and consider whether these words *to practise, to meditate* and *to love,* speaking of the precepts already given, signify anything else than to apply our whole soul to the practice of virtues. For our soul, being a Reasonable Substance, has necessarily three faculties: the first is that by which we learn, and is the faculty which is commanded to *meditate*; the second is that whereby we retain what we learn and put it in practice, and is the faculty which is required *to practise and to exercise*; and the third is that by which we love what we have learnt and what we practise, and is the faculty which is exhorted *to love* all these things.

In order, therefore, that all the faculties of our Rational Soul may apply themselves to those precepts of virtue and be wholly intent on them, of the Intelligent faculty he requires meditation, of the active, practice and exercise, and of the faculty that loves, he demands love, that by their means we may acquire the things that are truly good, that we may preserve them by exercise and always have for them an innate love in our hearts. Such a disposition as this never fails to be attended by divine hope, which makes the splendour of truth as eminently conspicuous in our souls as he himself promises us when he says, *It is they that will put thee in the way of Divine Virtue,* that is to say, they will make thee like God by the certain knowledge of all Beings. For the knowledge of the Causes of Being (I say, of the Causes) which are originally in the Intelligence of God, their Creator, *as Eternal Exemplars,* leads us to the sublimest pitch of the knowledge of God, which is followed by a perfect resemblance with Him. This is that Resemblance which is here called *Divine Virtue,* as being much superior to human virtue that precedes it and is, as it were, the foundation of it.

The first part of these Verses concludes with the love of Philosophy and of whatever is great and excellent. This love is followed by the knowledge of truth, and this truth leads us to a perfect resemblance with the Divine Virtue, as we shall see in the ensuing discourse.

The necessity of the union and of the connection of all these things is here confirmed by oaths, for the Poet swears fervently that the perfection of human virtue leads us to likeness with God.

As to the precept he gave us at the beginning of the Verses, *Reverence the Oath,* he commands us thereby to forbear swearing in casual things whose event is uncertain, for such things are of small moment and subject to change; therefore it is neither just nor safe to swear concerning them. But concerning the things here spoken of whose connection is fixed by Necessity and whose consequence is very great, we may swear safely and with all manner of decency and justice. For neither their instability will deceive us (since being linked by the Law of Necessity they cannot but arrive), nor their meanness and obscurity render them unworthy to be confirmed by the testimony and intervention of the Divinity. If virtue and truth are found in men, much more are they visible in the Gods.

Moreover, this *Oath* is in this place a precept that we ought to honour him who instructs us in the truth, so far as even to swear by him — if it be necessary for the confirmation of his doctrine — and not to say barely of him "he said it", but to assert with confidence, "the things are thus, I swear it by himself".

By swearing concerning the necessary connection and union of these most perfect habitudes, he enters into the very foundation of theology and manifestly demonstrates that the Tetraktys, or Number of Four, which is the source of the Eternal Order of the world, is nothing else than God Himself who has created all things.

But how comes God to be the Tetraktys? This thou may'st learn in the holy book that is ascribed to Pythagoras and in which God is celebrated as the Number of Numbers. For if all things exist by His eternal decrees, it is evident that in each species of things the number depends on the cause that produced it. There we find the first number, and thence it is come to us.

Now the finite interval of number is ten, for he who would reckon more after ten comes back to one, two, three, until by adding the second decad he makes twenty, by adding the third decad in like manner he makes thirty, and so goes on by tens until he comes to a hundred. After a hundred he comes back again to one, two, three, and thus the interval of ten always repeated will amount to an infinity.

The power of ten is four, for before we come to a complete and perfect decad we discover all the virtue and all the perfection of the ten in the four. For example, in assembling all the numbers from one to four inclusively, the whole composition makes ten, since one, two, three and four are ten. Four is an arithmetical middle between one and seven, for it exceeds the number one as much as it is exceeded by the number seven, and this number is three, four being as many more than one as seven is than four.

The powers and properties of the unit and of the septenary are very great and excellent, for the unit, as the principle of all the numbers, contains in itself the powers of them all. The seven, being a virgin and without any mother, holds in the second place the virtue and the perfection of the unit because it is not engendered by any number within the interval of ten, as four is produced by twice two, six by twice three, and eight by twice four, nine by three times three, and ten by twice five. Nor does it produce any number within that interval as the number two produces four, the three nine, and the five ten. And the four, holding the middle place between the uncreated unit and the motherless seven, has alone received the virtues and powers of the numbers producing and produced which are contained in the decad, being produced by a certain number and producing likewise another, for two being doubled begets four, and four being doubled begets eight.

Add to this that the first solid body is found in the Tetraktys, for a point answers to a unit, and a line to a binary, because indeed from one point we go to another point and this makes the line. The superficies answers to the ternary, for a triangle is the most plain of all rectilineal figures. But solidity is the nature of the Tetraktys, for it is in the four that we discover the first Pyramid, whose triangular basis is composed by the three and its point or top is made by the unit.

Moreover, there are four faculties that judge of things: understanding, knowledge, opinion and sense, for all things fall under the judgement of one of these four faculties. In a word, the Tetraktys contains and binds together all beginnings whatsoever, the elements, numbers, seasons, ages, societies or communities, and it is impossible to name any one single thing that depends not on that number as on its roots and principle. For, as we said before, the Tetraktys is the Creator and the Cause of all things.

The Intelligible God is the Cause of the Heavenly and Sensible God. The knowledge of this God was transmitted to the Pythagoreans by Pythagoras himself, by whom the author of these Verses swears in this place that the perfection of virtue will lead us to the light of truth. We may safely say that this precept, *Reverence the Oath,* is particularly observed in regard to the Eternal Gods who are always the same, and that in this place the Poet swears by him who taught us the Quaternary Number, who indeed was not one of these Gods nor of the *Heroes* who are such by nature, but only a man adorned with the likeness of God, who preserved in the minds of his disciples all the majesty of that Divine Image.

For this reason the Poet, in affairs of so great moment, swears by him, thereby tacitly to insinuate the great veneration his disciples had for him and the vast respect and esteem which this

Philosopher had acquired on account of the doctrine he taught.

The chief of his precepts was the knowledge of the Tetraktys that created all things. But seeing that the first part of this Verse has been briefly explained and that the latter part of it constitutes a firm and solemn promise that the sacred name of the Tetraktys is known by a hope that cannot deceive us, and seeing besides that this Divine Tetraktys has been explained as fully as the bounds which we prescribed to ourselves would allow, let us proceed to the other things to which these Verses summon us. But let us first show with what ardour and with what preparation we ought to apply ourselves thereto, and what need we have to be therein assisted and succoured by the Supreme Beings.

VERSES 48 AND 49

Never set thy hand to the work, till thou hast first prayed the Gods to accomplish what thou art going to begin.

The author of these Verses describes in a few words the two things that absolutely must concur to make us obtain the true goods. These two things are the voluntary motion of our soul and the assistance of Heaven. Though the choice of good be free and depends on us, seeing that we hold this liberty and this power from God, we have continual need of the assistance of God to cooperate with us and to accomplish what we ask of Him. For our own endeavours are properly like an open hand, stretched out to receive good things, and what He contributes on His part is like the storehouse or source of the gifts which He bestows upon us. Our part is to seek after that which is good, and it is the part of God to show it to them who seek after it as they ought. Prayer is a medium between our seeking and the gift of God. It is addressed to the Cause that has produced us and which, as it gave us our being, gave us our well-being likewise.

Now how can man receive any good unless God bestows it? And how shall God, who can alone give it, give it to him who, being the master of his own desires, disdains even to ask for it? In order, therefore, that on the one hand we may not pray in words only, but confirm our prayers by our actions, and that on the other we may not entirely trust, but implore the assistance of God on them and thus join our prayers and actions together as form to matter, the Poet asks us to pray for what we do and to do what we pray for, and so joins these two together and says: *Never set thy hand to the work, till thou hast first prayed the Gods to accomplish what thou art going to begin. . . .*

VERSES 49, 50 AND 51

When thou hast made this habit familiar to thee, thou wilt know the constitution of the Immortal Gods and of men; even how far the different Beings extend, and what contains and binds them together.

The first thing the author promises to those who practise the foregoing precepts is the knowledge of the Gods, the science of theology, and the ability to distinguish aright between all the Beings that flow from the sacred Tetraktys, along with their difference according to their kinds and their union in the order of the constitution of this Universe (for their order and their rank are in this place expressed by the word *constitution*). *How far the different Beings extend themselves* expresses their specific difference, and *what contains and binds them together* marks their generic community.

The several kinds of Rational Substances, though they are divided by their nature, reunite themselves by the same interval that divides them. The fact that some of them are first, others in the middle, and others last, is what at the same time separates and unites them, for by this means the first can neither be middle nor last, nor the middle first nor last, nor the last middle nor first, but they remain eternally distinguished and divided according to their genus, by the bounds which their Creator has prescribed for them. . . .

All these different Beings serve together to the perfection of the whole fabric, and this is what is here meant by these words: *and what contains and binds them together.*

Inasmuch as they differ in their kinds, they are separate from one another. But inasmuch as they are members of one and the same whole, they rejoin and reunite themselves. By this separation and this union together they complete and perfect the whole constitution and order of this Divine Work, a constitution that thou wilt know perfectly if thou would perfect the virtues of which he has already spoken.

We cannot mention the two extremes, but the middle will immediately present itself to the mind; therefore he thought it enough to say *the constitution of the Immortal Gods and of men.* For the first Beings are linked to the last by the middle Beings, and the last reascend to the first by the mediation and interposition of the *Heroes full of goodness and light.* This is the number and the order of Intelligent Beings, as we said in the beginning of this work, where we show that the first in this Universe are the *Immortal Gods,* after them the beneficent *Heroes,* and last of all the *Terrestrial Daimons,* whom he here calls *mortal men.*

How to know each of these kinds has been already shown in the beginning of this discourse, that is, by having a scientific knowledge of all these Beings which tradition has taught us to honour. This scientific knowledge is formed only in those who have adorned Practical Virtue with Contemplative Virtue, or whom the goodness of their nature has exalted from Human to Divine Virtues. Thus, to know the Beings as they were established and constituted by God Himself is to raise ourselves up to the Divine Likeness.

But as the order of these incorporeal or immaterial Beings succeeds the corporeal Nature, which fills this visible world and is submitted to the conduct of those Rational Essences, the Poet shows in the next place that the advantage of Natural or Physical Philosophy will be the effect of having learnt all these things in the order and by the method before mentioned.

VERSES 52 AND 53

Thou shalt likewise know, in accord with Cosmic Order, that the nature of this Universe is in all things alike, so that thou shalt not hope what thou oughtest not to hope; and nothing in this world shall be hid from thee.

Nature, in forming this Universe after the Divine Measure and Proportion, made it in all things conformable and like to itself, analogically in different manners. Of all the different species diffused throughout the whole, it made, as it were, an Image of the Divine Beauty, imparting variously to the copy the perfections of the original. To the Heavens it gave the perpetual motion, and to the earth stability. Now these two qualities are so many strokes and touches of the Divine Resemblance. He appointed the celestial body to surround the Universe, and the terrestrial body to serve as centre to the celestial.

Now in a sphere the centre and the circumference may be regarded in different respects, as its beginning and as its principle. Hence it is that the circumference is diversified with an infinity of stars and of Intelligent Beings, and that the earth is adorned with plants and with animals who are endowed only with sense.

Between these two sorts of Beings so different from each other, man holds the middle space as an amphibious animal, being the last of the superior Beings and the first of the inferior. This is the reason why he sometimes unites himself to the Immortal Beings, and by his return to understanding and to virtue recovers his natural state; and sometimes he falls back to the mortal, and by transgressing the Divine Law forfeits dignity. Indeed, being the last of Rational Beings, he cannot think and know always alike, for if he could he would not be man, but God, by nature; nor contemplate always, for that would place him in the rank of *Angels*. Whereas he is only man, who by resemblance and likeness may raise himself up to what is most good and excellent, and who by nature is inferior to the *Immortal Gods* and to the *Heroes full of goodness and light,* that is to say, to the two kinds that hold the first and second rank.

As he is inferior to these Beings inasmuch as he is not always engaged in meditation on them, but is sometimes in a total ignorance and forgetfulness of his own Essence and of the Light that descends from God upon him, so likewise when he recovers this forgotten knowledge, he is superior to all the animals without Reason and to plants. And he surpasses, by his Essence, all terrestrial and mortal Nature because he is himself naturally able to return to his God, to efface his forgetfulness by reminiscence, to recover by instruction what he has lost, and to repair his flight from the things above by a contrary tendency, that is to say, by being wholly intent upon them.

This being, therefore, the nature of man, it becomes him to know the constitution of the *Immortal Gods* and of mortal men (that is to say, the order and the rank of Rational Beings); to know that the nature of this Universe is in all respects alike (that is to say, that the whole Corporeal Substance, from the highest to the lowest, is honoured with an analogical likeness of God); and lastly, to know all these things *in accord with Cosmic Order,* that is to say, as they are established by the Law, as God created them, and in what manner the incorporeal as well as the corporeal are disposed and placed by His Laws. For this precept that commands us to know them *in accord with Cosmic Order* ought to be understood of both these two works of God. . . .

VERSES 54, 55, 56, 57, 58, 59 AND 60

Thou wilt likewise know that men draw upon themselves their own misfortunes, voluntarily and of their own free choice.

Wretches that they are! They neither see nor understand that their good is near them. There are very few of them who know how to deliver themselves out of their misfortunes.

Such is the Fate that blinds mankind and takes away his senses. Like huge cylinders, they roll to and fro, always oppressed by ills without number; for fatal contention, which is innate in them, pursues them everywhere, tosses them up and down, nor do they perceive it.

Instead of provoking and stirring it up, they ought by yielding to avert it.

The order of the corporeal and incorporeal Essences being well understood, we must necessarily comprehend the Essence of man and know what it is, to what passions it is subject, and that it holds the middle rank between the Beings that never fall into vice and the Beings that can never raise themselves up to virtue. Hence it is that it has the two tendencies which these two affinities naturally inspire, and lives sometimes an intellectual life and sometimes embraces affections that are wholly sensual.

This made Heraclitus say, with great reason, that our life is death, and our death life, for man falls and exiles himself from the Mansions of the Blest, as Empedocles the Pythagorean says:

Banished the blissful bowers,
Forlorn he wanders, by dire discord tossed,
And in impetuous storms of raging strife is lost.

But he reascends and recovers his primordial habitude if he scorn the things here below and detest this dismal abode where, as the same poet says, inhabit:

> Murder and rage, and a thousand swarms of woes.

And in which they who fall,

> Wander bewildered, helpless of relief,
> In the dark plains of injury and grief.

He who shuns these dismal plains of injury is led by that good desire into the meadow of truth. If he forsakes it his wings flag and fail him, and down he drops headlong into an earthly body,

> Where in large draughts
> He quaffs the oblivion of his happiness.

And herein agrees the opinion of Plato who, speaking of this fall of the soul, says: "But when, having no longer a strength sufficient to follow good, she beholds not this Field of Truth, and being by some misfortune filled with vice and with forgetfulness, she grows dull and heavy, and being thus stupefied, she comes to lose her wings and to fall down upon the earth; then the Law sends her to animate a mortal body."

Concerning the return of the soul to the place from which she descended, the same Plato says: "The man, who by his Reason, has overcome the tumult and wild disorder that are occasioned in him by the mixture of earth, water, air and fire, retakes this primitive form, and recovers his original habitude, because he returns sound and whole to the Star that had been assigned to him." He returns 'sound' because he is freed from the passions which are as so many diseases, and this cure cannot be performed in him except by the means of Practical Virtues. He returns 'whole' because he recovers understanding and knowledge as his essential attributes, which cannot happen to him but by the means of the Contemplative Virtues.

Moreover, the same Plato expressly teaches that only by our aversion for the things below can we heal and correct the apostasy that makes us go astray from God. He asserts that this avoiding of evils here below is Philosophy, only thereby insinuating that men alone are vulnerable to these sorts of passions, and that it is not possible that evils should be banished from the earth or that they can approach the Deity, but they hover always about the earth we dwell in and adhere to mortal nature as being the effects of unavoidable Necessity.

For the Beings that are subjected to generation and corruption may be affected and have desires contrary to Nature, and this is the principle of all evils. To teach us how to avoid them, Plato adds: "Therefore we ought to fly from hence with all diligence; now to fly from hence is to endeavour to resemble God as much as man is capable of doing; and to resemble God is to become just and holy with prudence." For he who would avoid these evils ought to begin by divesting himself of this mortal nature, it being impossible that they who are engaged in it should not be plunged in all the evils that Necessity produces therein.

In like manner, therefore, as our swerving and absence from God and loss of the wings that raised us up towards the things above have caused our exile in this region of death, the abode of all evils, so the divesting ourselves of all worldly affections and the renovation of virtues like new growth of our wings to guide us to the Mansion of Life — where true goods are to be found without the least alloy of evil — will bring us back to happiness. For the Essence of man, holding the middle place between the Beings that always contemplate God and those that are incapable of contemplating Him, may raise itself up towards the one, or debase and sink itself down towards the other, having by reason of its amphibious nature an equal propensity to take the Divine or brutal resemblance accordingly as it receives or rejects the understanding or the Good Spirit.

He therefore that knows this liberty and this double power in human nature knows likewise how men voluntarily draw on themselves their own evils and how they become wretched and miserable by their own election. For though they could have remained in their true country, they suffer themselves to be dragged to birth by the inordinateness of their own desires. And when they might readily free themselves from this miserable body, they voluntarily immerse themselves into all the confusions and into all the disorders of the passions. This is what the Poet would have us understand when he says: *They neither see nor understand that their good is near them.*

This good is *virtue* and *truth*, and *not to see that they are near* is to lack the inclination to search after them. *Not to hear or understand that they are near* is to fail both to hearken to the admonitions and to obey the precepts that others give. For there are two ways of receiving knowledge: one by instruction, as by the hearing, the other by search, as by the sight. Men are therefore said to draw evils upon themselves of their own accord when they will neither learn from others nor find out for themselves, remaining destitute of the sense of all good and thus entirely useless and unprofitable. For every man who does not know himself and does not hearken to the instructions of others is entirely useless and in a desperate condition. But they who endeavour to find themselves or to learn from others the things that are truly good are those of whom the Poet says that they *know how to deliver themselves out of their misfortunes,* and who, by avoiding the troubles and labours of this world, transport themselves into the pure and free *Aether.*

The number of these is very small, for far the greatest part of men are wicked slaves to their passion, and in a manner run mad through the violence of their propensity to the things of this world. This evil they bring upon themselves by having wilfully departed from God and deprived themselves of His Presence and, if I may dare to say so, of the familiarity with Him which they had the happiness to enjoy while they inhabited the Mansions of pure and unclouded Light. Now the *Fate that blinds mankind and takes away his senses* is a mark of their departure from God. . . .

Since the cause of this *Fate* that takes from men their understanding of God is the abuse they make of their freedom, he teaches in the two following Verses how to reform this abuse and how to make use of the same freedom to return to God.

In order to prove to us that we draw on ourselves our own evils only because we will have it so, he says *fatal contention, which is innate in them, pursues them everywhere, tosses them up and down, nor do they perceive it.* And immediately after, to show that the remedy is in our own hands, he adds, *instead of provoking and stirring it up, they ought by yielding to avert it.* But perceiving at the same time that we have, before all things, need of the assistance of God to enable us to depart from evil and to embrace good, he adds forthwith a sort of prayer and makes an ejaculation to God, the sole means to procure His assistance.

VERSES 61, 62, 63, 64, 65 AND 66

Great Jupiter, Father of men, you would deliver them all from the evils that oppress them, if you would show them what is the Daimon of whom they make use.

But take courage, the race of men is divine. Sacred Nature reveals to them the most hidden Mysteries.

If she impart to thee her secrets, thou wilt easily perform all the things which I have ordained thee, and healing thy soul, thou wilt deliver it from all these evils, from all these afflictions.

It was the custom of the Pythagoreans to call God, the Father and Creator of the Universe, by the name of *Jupiter*, which in the original tongue is taken from a word that signifies 'Life', for He who gave Life and Being to all things ought to be called by a name derived from His power. And the truly proper name for God is that which most evidently denotes His works.

In our day, we may much rather say that custom and common agreement have produced the names that seem to us most proper, than that the propriety of their nature gave occasion to their invention. This is evident from a world of names that are given to things contrary to the very nature of them, and with which they agree no more than if we should call a wicked man good or an impious man pious. These sorts of names have not the conformity and suitableness that names ought to have, inasmuch as they denote neither the being nor the qualities of the thing on which they are imposed. But this agreement and this propriety of names ought most of all to be sought after in the things that are eternal, and amongst the eternal in the divine, and amongst the divine in the most excellent.

Thus you see why the name of *Jupiter* carries even in the very sound a symbol and an image of the Essence that created all things. For the first composers of names (like excellent statuaries) by their sublime knowledge and wisdom expressed by the names themselves (as by animated images) the virtues and qualities of those to whom they gave them. They invented names whose very sound was the symbol of their thoughts, and their thoughts were most resembling and most instructive images of the subjects on which they thought.

And indeed these great souls, by their continual application to Intelligible things — being as it were swallowed up in contemplation and grown, as I may say, pregnant with this commerce — when they were taken with the pangs of bringing forth their thoughts, cried out in expressions and gave such names to things as by their very sound and by the letters employed in forming them, they perfectly expressed the kinds of the things named, and led to the knowledge of their nature all who comprehended them aright. The end of their contemplation has been, in regard to us, the beginning of knowledge. Thus the Creator of all things was called by these men of deep knowledge and wisdom sometimes by the name of *Four*, and sometimes by the name of *Jupiter,* for the reasons which we have already mentioned.

Now what we ask of Him in this prayer is what He bestows on all men by reason of His infinite goodness, but it depends on us to receive what He is continually giving. It was said before, *Never set thy hand to the work, till thou hast first prayed the Gods to accomplish what thou art going to begin,* to teach us that the *Gods* are always ready to give us the things that are good, but that we receive them only when we ask for them and when we stretch out our hands to this divine distribution. For what is free receives not what is truly good unless it wills. And the true goods are truth and virtue which, flowing without ceasing from the Essence of the Creator, are visible at all times and in the same manner to the eyes of all men.

And when these Verses pray that we may be delivered from all our evils, they ask, as a thing absolutely necessary, that we may know our own Essence. For this is what is meant by this expression *what is the Daimon of whom they make use,* that is to say, *what is their soul.* For from this return to ourselves, from this knowledge of ourselves, will necessarily result the deliverance from our evils and the manifestation of the goods that God offers us, to make us happy. This Verse, therefore, supposes that if all men know what they are and *what is the Daimon of whom they make use,* they will all be delivered from their evils. But this is impossible, for it cannot be that they should all apply themselves to Philosophy or that they should receive equally all the good things that God incessantly offers for the perfection of happiness.

What, then, remains but for those only to take courage who apply themselves to the knowledge that alone discovers our true good, the good that is proper for us. They only will be delivered from the evils that are inherent in this mortal nature because it is they alone who addict themselves to the contemplation of the things that are truly good. Therefore they deserve to be placed amongst the number of the Divine Beings because they are instructed by Sacred Nature, that is to say, by Philosophy, and because they practise all the precepts which their duty obliges them to observe.

Now if we have any conversation with these divine men, we shall make it visible by applying ourselves wholly to good works and to the intellectual sciences, by which alone the soul is healed of its passions and delivered from all the evils here below, and thus is translated into an Order and a condition wholly divine. . . .

VERSES 67, 68 AND 69

Abstain thou from all that we have forbidden in the Purifications; and in the Deliverance of the Soul make a just distinction of them; examine all things well, leaving thyself always to be guided and directed by the understanding that comes from above, and that ought to hold the reins.

The Rational Essence, having received from God, its Creator, a body conformable to its nature, descended upon earth, so that it is neither a body nor without a body. But being incorporeal, it has nevertheless its form determined and bounded by the body, even as in the stars. For their superior part is an incorporeal Substance and their inferior a corporeal, the sun itself being a compound of something corporeal and something incorporeal. Not that it is composed of two parts which, having been separate, have united themselves together — for if so they might separate themselves again — but of two parts created together and born together with subordination, so that the one directs and the other obeys.

It is the same with all Rational Essences, as well with the *Heroes* as with men. For a *Hero* is a Rational Soul with a luminous body, and man is also a Rational Soul with an immortal body created with it. Thus you see the doctrine of Pythagoras, which Plato in his *Phaedrus* explained long after him, comparing the divine soul and the human soul to a winged chariot that has two horses and a coachman to guide it.

For the perfection, therefore, of the soul we have need of truth and virtue. And for the purgation of our luminous body we stand in need to be cleansed of all the pollutions of matter, to have recourse to holy Purifications, and to make use of all the strength that God has given us to stir us to fly from these inferior abodes. And this is what the preceding Verses teach us.

They instruct us to remove the pollution of matter by this precept, *Abstain from all that we have forbidden.*

They enjoin us to add to this abstinence holy Purifications and the strength with which we are divinely inspired. Such a command is indeed a little obscurely inculcated by these words: *in the Purifications and in the Deliverance of the Soul,* etc. They endeavour to render the form of the human Essence entire and perfect by adding, *leaving thyself always to be guided and directed by the understanding that comes from above, and that ought to hold the reins.* The Poet thereby sets before our eyes the whole human Essence and distinguishes the Order and the rank of the parts that compose it. That which guides is as the coachman, and that which follows and obeys is as the chariot.

These Verses, therefore, teach all who wish to understand the *Symbols* of Pythagoras that by the exercise of virtue, and by embracing truth and purity, we ought to take care of our soul and our luminous body, which the Oracles call the *subtle chariot of the soul.* Now the purity here spoken of extends to food and drink and to the whole management and usage of our mortal body, in which is lodged our luminous body which inspires life into the inanimate body and contains and preserves all its harmony.

For the immaterial body is life and produces the life of the material body. By this life our mortal body becomes perfect, being composed of the immaterial life and the material body and being the image of the whole *Man* who, properly speaking, is a compound of the Rational Essence and of the immaterial body.

Since man is composed of these two parts, it is evident that he ought to be purified and perfected in both of them. To this purpose he must follow the ways that are proper to each of his two natures, for each part requires a different purgation.

The Reasonable Soul, in regard to its faculties of reasoning and judging, must be purged by truth, which produces knowledge; with regard to its faculty of deliberating it must be purged by consultation that it may contemplate the things that are divine and regulate the things below. For the first we have need of truth and for the last we have need of virtue, that we may wholly apply ourselves to the contemplation of the things that are eternal and to the practice of all our duties. In both we shall avoid the storms that folly creates if we obey exactly the Divine Laws that have been given to us.

For this folly is the thing of which we ought to purge our Rational Essence. It was that very folly that gave it a tendency and affection for the things here below. But because a mortal body has annexed itself to our luminous body, we ought likewise to purge it of the corruptible body and deliver it from the sympathies it has contracted with it.

There remains, therefore, only the purgation of the spiritual body, which must be done according to the sacred Oracles and to the holy method which the art teaches. But this purgation is in some manner more corporeal, and therefore employs all sorts of ways to heal this vivifying body and to counsel it by this operation to separate itself from matter and to take its flight to those blissful regions where its happiness originated.

And whatever is done for the purgation of this body, if it be done in a manner worthy of God, without any deceit or imposture, will be found consonant with the rules of truth and virtue. For the purgations of the Reasonable Soul and the luminous chariot are performed to this purpose, that this chariot may come to have wings, and immediately soar aloft to its Celestial Habitations.

Now what most contributes to the growth of these wings is meditation, by which we learn little by little to wean our affections from earthly things and acquire the habit of contemplating the things that are immaterial and Intelligible and shake off the pollutions it has contracted by its union with the terrestrial and mortal body. And indeed, by these three aids it rouses itself up, it is filled with divine vigour and reunites itself to the Intelligent

Perfection of the soul.

But it will be asked wherein and how the abstaining from meat can contribute to things of such excellence. Certainly it cannot be doubted, especially in regard to those who are accustomed to avoid all worldly joy, that to abstain entirely from eating meats, especially from such as enervate the mind and incline the body to lustful desires, will be a great help to them and a considerable advance towards their purification.

This is the reason why in the symbolical precepts we are enjoined to observe these abstinences. Such precepts when considered in the mystical sense conceal one that is general and of large extent. Though at the same time literally taken, they have the sense which they present, and positively forbid what is expressly named in the precepts. For example, the symbol that forbids *us eat the matrix of any animal,* when literally taken, forbids the eating of a certain part of it, and that a very small one too. But if we dive to the bottom of the hidden sense of this Pythagorean mystery, we shall discover that by this palpable and sensible image we are taught entirely to renounce whatever relates to birth and generation. And as we are commanded to abstain actually, and according to the letter, from eating that part of any animal, so are we to practise and observe with equal care the most mysterious and hidden injunctions in these precepts in order to achieve the purgation of the luminous body.

In like manner in this precept, *Thou shalt not eat the heart,* the chief sense is that we should avoid anger. But the literal and subordinate sense is that we should abstain from eating that forbidden part.

So the design of the precept that commands us *to abstain from the flesh of beasts that die of themselves,* is not only to wean us in general from this mortal nature, but to hinder us likewise from partaking of any profane flesh and of such as is not fit for sacrifices. For in the symbolical precepts it is just to obey the literal as well as the hidden sense, and the practice of the literal sense is the only way to attain to the observance of the mystical, which is the more important.

Thus too we ought to understand this Verse which in two or three words gives us the seeds and principles of the best works. *Abstain from all that we have forbidden,* which is the same as if it had said *Abstain from mortal and corruptible bodies.* But because it is impossible to abstain from all, he adds, *that we have forbidden,* and he points out the places of which he speaks, *in the Purifications and in the Deliverance of the Soul,* to the end that by abstaining from prohibited meats we may increase the splendour of the corporeal chariot, and take such a care of it as becomes a soul that is purified and delivered from all the pollutions of matter.

He leaves the true distinction of all these things to the understanding which, being the only faculty that judges, is alone capable of taking such care of the luminous body as the purity of the soul requires.

Thus you see why he calls the understanding *the coachman, the conductor,* that holds the reins, since it has been created to guide the chariot. It is called the understanding because it is the Intelligent Faculty, and *conductor* or *coachman* because it governs and guides the body.

Now the Eye of Love is what directs the coachman, for though it be an Intelligent Soul, it is only by the assistance of the Eye of Love that it discovers the Field of Truth, and by the faculty that serves it, it curbs and restrains the body that is annexed to it and, guiding it with wisdom, becomes the master of it and turns it towards its own self, to the end that the whole composition may be entirely taken up with contemplating the Divinity and may conform itself wholly to its image.

This is, in general, the abstinence here spoken of, and all the great goods to which it endeavours to lead us. All these things are particularly delivered in the holy precepts that have been given us under shadows and under veils. And though each of them ordains a particular abstinence — as from beans among the legumes, among beasts from the flesh of such as die of themselves (though the very kinds be expressed, as *Thou shalt not eat the gurnet* for the fish, nor such an animal for the beasts, nor such a bird for the fowls of the air) — nevertheless in each of these precepts the author includes all the perfection of purification. For though he literally forbids such and such a thing, as to bodily abstinence, by reason of certain physical properties and virtues, yet in each precept he insinuates our purgation from *all* carnal affections, and teaches man to return home to himself that he may bid farewell to this abode of generation and corruption and take his flight to the Elysian Fields and to the most pure Aether. . . .

Now by mystical operation I mean the purgative faculty of the luminous body, to the end that of all Philosophy the theory may precede as the Mind, and the practical follow as the Act or Faculty. Now the practical is of two sorts, political or civil and mystical. The first purges us of folly by the means of virtue, and the second cuts off all earthly thoughts by means of the sacred ceremonies.

The public laws are a good pattern of civil Philosophy and the sacrifices offered by cities of the mystical. Now the sublimest pitch of all Philosophy is the contemplative mind; the political mind holds the middle place; and in the last is the mystical. The first, in regard to the two others, holds the place of the eye; and the two last, in regard to the first, hold the place of the hand and of the foot. But they are all three so well linked together that either of the three is imperfect and almost useless without the cooperation of the other two. Therefore we ought always to join together the knowledge that has found out the truth, the faculty that produces virtue, and that which brings forth purity, to the end that the civil actions may be rendered conformable to the mind that presides, and that the holy actions may be answerable to the one and to the other.

Thus you see the end of the Pythagorean Philosophy is that we may become all over wings to soar aloft to the Divine Good, to the end that at the hour of death, leaving upon earth this mortal body and divesting us of its corruptible nature, we may be ready for the celestial voyage, like champions in the sacred combats of Philosophy. For then we shall return to our ancient country and be deified as far as it is possible for men to become *Gods*. And this we are promised in the two following Verses.

VERSES 70 AND 71

And when, after having divested thyself of thy mortal body, thou arrivest in the most pure Aether, thou shalt be a God, immortal, incorruptible, and death shall have no more dominion over thee.

Behold the most glorious end of all our labours! Behold, as

Plato says, the glorious combat and the great hope that is proposed to us! Behold the most perfect fruit of Philosophy! This is the greatest work, the most excellent achievement of the Art of Love, that mysterious Art which raises all souls to Divine Goods and establishes them therein and delivers them from afflictions here below, as from the obscure dungeon of mortal life. It exalts to the Celestial Splendours and places in the Islands of the Blest all who have walked in the ways which the foregoing rules have taught them. For them and them alone is reserved the inestimable reward of deification, it not being permitted for any to be adopted into the rank of the *Gods*, but for him alone who has acquired for his soul virtue and truth, and for his spiritual chariot, purity.

Such a man, having thereby become sound and whole, is restored to his primordial state after he has recovered himself by his union with sound Reason, after he has discovered the All-Divine Ornament of this Universe and thus found out the Author and Creator of all things, as much as it is possible for us to find Him. Having arrived after purification to that sublime degree of bliss which the Beings whose nature is incapable of descending into generation always enjoy, he unites himself by his knowledge to this *Whole*, and raises himself up even to God Himself.

But since he has a body that was created with him, he stands in need of a place wherein he may be seated, as it were, in the rank of the stars. And the most suitable place for a body of such a nature is immediately beneath the moon, as being above all terrestrial and corruptible bodies, and beneath all the Celestial. And this place the Pythagoreans call the *pure Aether* — *Aether* because it is immaterial and eternal, and *pure* because it is exempt from earthly passions.

What shall he be, then, who is arrived there? He shall be what these Verses promise him, *an Immortal God*. He shall be rendered like the *Immortal Gods* of whom we have spoken in the beginning of this treatise, an *Immortal God*, I say, but not by nature. For how can it be that he who since a certain time only has made any progress in virtue, and whose deification has had a beginning, should become equal to the *Gods* who have been *Gods* from all eternity? This is impossible. Therefore, to make this exception and to mark this difference, the Poet, after he had said *Thou shalt be a God*, adds, *immortal, incorruptible, and death shall have no more dominion over thee,* thereby intimating that it is a deification which proceeds only from our being divested of what is mortal and is not a privilege annexed to our nature and to our Essence, but to which we arrive little by little and by degrees. So this class of Beings makes a third sort of *Gods* who are immortal when they are ascended into Heaven but mortal when they descend upon Earth, and in this way always inferior to the *Heroes who are full of goodness and light.* These last remember God always, but the third sometimes forget Him. For it is not possible that the third kind, though rendered perfect, should ever be superior to the second or equal to the first. But continuing always as the third, they become like the first though they are subordinate to the second, for the resemblance to the *Celestial Gods* is more perfect and more natural in the Beings of the second rank, that is to say, in the *Heroes.*

Thus there is but one and the same perfection common to all Intelligent Beings, which is their resemblance to God who created them. But see what makes the difference: this perfection is always, and always the same, in the Celestial; always too, but not always the same, in the Aethereal *(Heroes)* who are fixed and permanent in their state and condition; and neither always, nor always the same, in the Aethereal (souls of men) who are subject to descend and to come and inhabit the earth.

If any man should assert that the first and most perfect likeness of God is *the copy and the original of the two others,* or that the second is of the third, his assertion would be very just. Our aim is not only to resemble God, but to resemble Him by approaching the nearest we can to this all-perfect Original, or to arrive at the second resemblance. But if, not being able to attain to this most perfect resemblance, we acquire that of which we are capable, we have, as well as the most perfect Beings, all that our nature requires. And we enjoy the perfect fruits of virtue even in this, that we know the measure and extent of our Essence and that we are not dissatisfied with it.

For the perfection of virtue is to keep ourselves within the limits of the Creation, by which all things are distinguished according to their kind, and to submit ourselves to the Laws of Providence that have distributed to each individual the good that is proper for it in regard to its faculties and its virtues.

This is the commentary we have thought fit to make on these *Golden Verses* and that may be called a summary, neither too prolix nor too succinct, of the doctrine of Pythagoras. It was not fitting either that our explanation should imitate the brevity of the text, for then we should have left many things obscure and should not have been able to discover and show the reasons and the beauties of all the precepts, or that it should contain all his Philosophy, for that would have been too large and too tedious a work for a commentary. But we thought it proper to proportion this work, as much as we could, to the sense of these Verses, reciting no more of the general precepts of Pythagoras than what was consonant and might serve to the explanation of these *Golden Verses,* which are properly only a most perfect representation of his Philosophy, an abridgement of his principal tenets and elements of perfection, which they who have walked in the ways of God and whose virtues have raised up to Heaven have left to instruct their descendants. These elements may justly be called the greatest and most excellent mark of the nobility of man, and are not the private opinion of any particular person but the doctrine of the whole sacred body of the Pythagoreans and, as it were, the common voice of all their assemblies. For this reason there was a law which enjoined each of them, every morning when he rose, and every night at his going to bed, to have these Verses read to him as the Oracles of the Pythagorean doctrine, to the end that by continual meditation on these precepts their spirit and energy might shine forth in his life. And this is what we likewise ought to do, that we may make trial, and find what great advantages we should in time gain by so doing.

(Adapted from the translation of N. ROWE)

OM

GLOSSARY

Acousticoi	Listeners; name given to those in the probationary degree of the School of Pythagoras
Aether	The immaterial and eternal substratum surrounding the earth in which pure intelligences have their abode; according to Hierocles, *aether* is restricted to the sublunary realm, but in some usages it applies also to the celestial spheres
Anamnesis	Recollection; in Pythagorean and Platonic philosophy, recollection of knowledge innate to the immortal soul
Cosmos	The universe as an intelligent and intelligible, ordered world
Daimon	The inner voice of conscience and intuition; an aspect of the human soul; Hierocles identifies terrestrial *daimones* with human souls
Hieros Logos	Sacred discourse; divine reason, word or cause
Magna Graecia	Greek colonies in Italy and Sicily
Nemesis	The precipitation of the consequences of past errors; the inexorable conditions of inevitable downfall; personified as a goddess
Oath	The solemn affirmation of a binding vow, thus sustaining divine order; the Pythagorean Oath was taken in the name of the Holy Tetraktys
Shravaka	Listener; the first step in discipleship; *cf. acousticoi*
Tetraktys	The sacred Quaternion; the Number of numbers; the Source of Nature; manifest Deity; the creative principle, represented by the triangle containing ten points in four rows, symbolizing the creative triad, the manifesting tetrad and the basic integers (1 + 2 + 3 + 4 = 10), as well as the point (1), line (2), figure (triangle, 3) and solid (tetrahedron, 4)

CONCORD GROVE PRESS

The CGP emblem identifies this book as a production of Concord Grove Press, publishers since 1975 of books and pamphlets of enduring value in a format based upon the Golden Ratio. This volume was typeset in Journal Roman Bold and Bodoni Bold, printed on acid-free paper and Smyth sewn. A list of publications can be obtained from Concord Grove Press, 1407 Chapala Street, Santa Barbara, California 93101, U.S.A.